THE OMEGA CONSPIRACY

DEFENDER
Crane, MO

The Omega Conspiracy:
Satan's Last Assault on God's Kingdom

Defender Publishing, Crane 65633

Published 2008 Printed in the United States of America.

ISBN-10: 1948014874 (paper)
EAN-13: 9781948014878 (paper)

A CIP catalog record for this book is available
from the Library of Congress

Cover illustration and design by Jeffery Mardis

Contents

Ω

Foreword

I first met Dr. I. D. E. Thomas in a restaurant in Pasadena, California. It was around 1992, and I was excited about meeting the man who had written *The Omega Conspiracy* and looked forward to getting my dog-eared, highlighted, notated copy of his book autographed. He was waiting for me in the lobby—slight of frame, graying hair, and an endearing, enigmatic smile that welcomed me. We shook hands and settled in our seats, as I quickly pulled out my tape recorder and started the interview. It was a lively exchange of questions that focused on the sons of God, the Nephilim, the first century Jewish historian Josephus, and how it all linked together with the ongoing UFO phenomena. What a ride!

The Omega Conspiracy had set the bar for a Christian response to the ongoing phenomena of UFOs. Remember this was all before movies like *Independence Day* or shows like the *X-Files*, which would help in the enculturation of UFOs.

The Omega Conspiracy is a masterful work of erudition and scholarly application that leaves the reader with concise information about spiritual phenomena manifesting on our planet today. It is a must read, a starting place, for anyone interested in the burgeoning phenomena of UFOs.

Many have followed in Dr. Thomas's footsteps, and we all owe him a great debt of gratitude for having the courage to set in print a subject that many in the Christian community would not even broach. He truly is *the* pioneer in this regard.

The times in which we live are tumultuous and uncertain. *The Omega Conspiracy* provides us with much-needed answers to the mystery of the sons of God. In my opinion we are in the days that Jesus warns of in Matthew 24:37: *As in the days of Noah, so it will be when the son of man returns.* This begs the question, what differentiates the days of Noah from any other? The answer is the presence of the sons of God and the *Nephilim* manifesting on the earth.

It is an honor for me to write this foreword. I salute with the utmost respect, admiration, and thanksgiving Dr. Thomas for his groundbreaking work and also Anomalos Publishing for republishing this wonderful tome.

Dr. L. A. Marzulli
Fall 2007

Introduction

Before long there will be heard throughout
the planet a formidable cry, rising like the howling
of innumerable dogs to the stars,
asking for someone or something to take command...

—Ortega y Gasset

A thousand years ago men believed that the Earth was flat. Anyone who ventured westward out of the Mediterranean Sea, beyond the "Pillars of Hercules," would inevitably fall off the edge of the world. They chiseled on the Rock of Gibraltar "Ne plus ultera" (No more beyond). But one day some adventurous seafaring men did sail beyond the Straits of Gibraltar, and they discovered that there was more beyond.

Five hundred years ago men believed that Earth was the center of the Universe. The sun, moon, and stars all circled the Earth, rising in the east and setting in the west. The Nicolaus Copernicus, the Prussian astronomer, wrote his "Do Revolutionibus Orbius Coelestium," proving that men before him had been wrong.

Two hundred years ago, the famous Lavoiser with a committee of the French Academy decided that "it was impossible for stones to fall from the sky because there are no stones in the sky." They thus concluded that stones only appeared to fall. Meteorites, they argued, were merely stones on Earth that had been struck by lightning.

A hundred years ago, just after Thomas Edison had invented his carbon filament lamp, a Parliamentary Committee of Inquiry was set up in England to report on the matter. Its chairman, Sir William England Preece, announced the group's findings to the House of Commons. He said that "electric light in the home was fanciful and absurd."

Less than a hundred years ago, men believed that man could never fly. Bishop Wright said that flying was "reserved for the angels." Then his two sons, Wilbur and Orville, came and proved their father wrong.

Less than fifty years ago, Sir Harold Spencer Jones, the director of Greenwich Observatory, announced that man would never set foot on the surface of the moon. Nevertheless, in July 1969 Neil Armstrong did exactly that.

Twenty-five years ago, Dr. Richard Woolley, Britain's Astronomer Royal, said that "space travel is utter binge." Since then American and Russian astronauts have proven him wrong.

Less than twenty-five years ago, men believed that the stories of UFOs circling the globe and landing on this planet were hoaxes or hallucinations. Yet more than 15 million witnesses equipped with photographs, movies and stills, physical records and scientific data have proved again that men can be wrong.

The Spiritual Nebulae

From the beginning, mankind has been mystified by the sky. But it is a fascination tinged with awe. The sky speaks to us of sheer vastness, fathomlessness, otherness…The spiral nebulae seems to know no limits. Astronomers, equipped with giant and sophisticated telescopes, have probed to a distance of 1,000,000 billion, billion miles…and still it is not the end. Puny man stands awe-struck and silent before such awesome proportions.

Although endless, the skies are not empty. They are populated with innumerable planets. Suns and stars, galaxies and constellations outnumber the grains of sand on the seashores of the world. But the skies are also the abode of the Great and High God. Focusing heaven-ward, we come away conscious of His greatness and of our littleness.

This attitude toward space is characteristic of man everywhere. "The idea of which regards the sky as the abode of the Supreme Being, or as identical with Him, is as universal amongst mankind as any religious belief can be, and is traceable back to the most primitive stages of culture known to us."[1]

To no one is this mystery and fascination more so than to the Christian believer.

Signs from the Heavens

The first verse of Scripture points him to the skies: "In the beginning God created the heavens and the earth" (Genesis 1:1). But it is only the beginning. Later on we read: "The heavens declare the glory of God, and the firmament sheweth forth His handiwork" (Psalms 19:1). And: "It is He that sitteth upon the circle of the earth, and the inhabitants thereof are as grasshoppers, that stretcheth out the heavens as a curtain, and spreadeth then out as a tent to dwell in…To whom will ye liken Me, or shall I be equal? Saith the Holy One. Lift up your eyes on high, and behold who hath created these things, and called them all by names by the greatness of His might, for that He is strong in power; not one faileth" (Isaiah 40:22, 25, 26).

Not only have men of the past turned toward the stars, but also men of the future will do so even more. The prophet Joel placed it on record: "And I will show wonders in the heavens and in the earth, blood and fire, and pillars of smoke" (Joel 2:30).

Jesus Himself said: "Nation shall rise against nation, and kingdom against kingdom: And great pestilences; and fearful sights and great signs shall there be from **heaven**" (Luke 21:10–11).

Other Devotees

Not only Christians look to the skies, the devotees of astrology are doing the same on an unprecedented scale. These people plan their lives with the planets. They believe that masses of rock and dust whirling frantically in space have a message for every individual on Earth, as well as for the planet itself. And it is with a rare gift of ingenuity that they clothe their suppositions with theological and biblical terminology.

Four thousand years ago, astrologers tell us, it was the Age or Aries. This was the time of the Old Testament, and what greater symbol could there be for that period than Aries the Ram. The God of the Old Testament was the Good Shepherd, and all the great celebrities—Abraham, Moses, David, were shepherds.

Two thousand years ago, the astrologers claim, it was the Age of Pisces. What better symbolism could there be for the New Testament? Pisces means "fish." Most of the disciples were fishermen, and Christ called them to be "fishers of men." The sign of the fish became their symbol.

Today, we are supposed to have entered the Age of Aquarius. Aquarius is the sign of the water carrier or the gardener. This, astrologers assure us, will be the age of planting seeds.

While it is true that the stars of heaven were created for signs, and that Christ foretold that special signs would appear in the heavens in the last days, this is a far cry from the artificial tenets of astrology. The phenomena of which Christ spoke had nothing to do with horoscopes, Zodiac influences, or predictions of human foibles and fancies. Christ spoke of signals that would warn men of the end of the age. To the unbeliever these were of catastrophe and calamity, to the believer, his final redemption. "Look up, and lift your heads; for your redemption draweth nigh" (Luke 21:28). It is from these same skies that Jesus will come back to us: "This same Jesus, which has taken up from you into heaven, shall so come in like manner as ye have seen

Him go into Heaven" (Acts 1:11). Millions of Christians throughout the ages have reechoed the sentiments of hymn-writer, W. Y. Fullerton:

> But this I know, the skies will thrill with rapture,
> And myriad, myriad human voices singing,
> And earth to heaven and heaven to earth, will answer,
> At last the Saviour, Saviour of the world, is King!"

Extraterrestrials Return

But this is not all. To our generation, more than to any other, the heavens have revealed yet another strange, frightening, fascinating phenomenon. In unmistakable terms they tell us that we are not alone in the universe. Other beings exist out there, and they are coming here. No longer is this something to fuel our imagination or titillate our curiosity. Rather, it is something to engage our most serious and rational attention.

If Ufology has performed no other service than this, it is commendable for again turning man's attention to the skies. It is from up there that our help will come for, the Bible declares, salvation from above.

So far so good. But the question is: Does our salvation lie with visitors from outer space? Is our redemption in the hands of extraterrestrial who supposedly man and operate UFOs? Are the creatures who sired the Nephilim in the days of Genesis returning as "Saviours from the Skies" or "Brethren from Space"? This book explains the phenomenon of Unidentified Flying Objects and offers an explanation that could identify the "beings" who operate them.

When this study began, the idea of a super race called Nephilim living on this planet seemed an incredible proposition. To proceed and predict their return was even more preposterous. But as my research progressed, the evidence became overwhelming. The return of the Nephilim, a super race sired by beings from another dimension, is the only viable explanation for what is happening. As incredible as this sounds, let us regard the ancient saying of Heracleitus who, 500 years before Christ said, "Because it is sometimes so unbelievable the truth escapes becoming known."

In some matters concerning creatures from outer space we can only speculate and theorize. But the Bible has much to say on the subject. What exactly is the biblical evidence for the existence of such beings? Does it bring these entities known as the Nephilim within the range of credibility? Not only credibility but I believe it supplies historical documentation. It offers amazing insights into their existence, origin, and identity and warns of the possibility of their re-appearance in the end-times.

This book is a thorough examination of this fascinating phenomena and how it affects us today.

1

Cosmic Riddles

About the Time of the End, a body of men will be raised up who will turn their attention to the Prophecies, and insist upon their literal interpretation, in the midst of much clamor and opposition.

—Sir Isaac Newton (1642–1727)

Twentieth century man is still at a loss to explain the amazing knowledge and expertise that characterized some of his ancient predecessors.

Unlike primitive savages roaming wild and naked in the bush, they planned the pyramids, built Babylon, engineered Stonehenge, and structured the Mayan Caracol.

A thousand years before Darwin formed his theory of evolution, ancient Mayans and Toltecs carved their own version of evolution in stone; their calendar was more precise than ours today, they even knew and used penicillin; and like the ancient Egyptians they designed great cities and mammoth pyramids. But from where, or whom, did they get their information and expertise?

Where did the ancient Dogon of the West African Republic of Mali receive their knowledge of Siriu's *invisible* satellite? Sirius, a star of the first magnitude in the constellation Canis Major, is 8.5 light years away, yet they knew the satellite's position, gravitation, and orbit. The star's white satellite was not discovered until 1844, and was not seen by telescope until 1862. But the Dogons knew of it long before the telescope.

Why did the Assyrians who lived more than 2,000 BC encircle the planet Saturn with a ring of serpents? How could they know of Saturn's rings? They depicted no other planet in this way. In addition to encircling Saturn with Serpents, they also recorded the different phases of the moon with an accuracy not seen again until the seventeenth century AD. But they did it in 1440 BC!

How did those ancient Greeks know that there were seven stars in Pleiades? They could only see six! Did a higher intelligence inform them?

Where did the Sumerians of Mesopotamia gain their expertise? According to researchers Alan and Sally Lansburg, "They pop up like some devilish jack-in-the-box, around 3000 BC, fully equipped with the first written language, sophisticated mathematics, a knowledge of physics, chemistry, and medicine."[2]

The pre-Inca mountaineers of Peru performed amputations, bone transplants, cauterizations, brain surgeries, and a variety of other complicated operations. From whom did they learn their advanced surgical skills?

From whom did builders of the awesome structures of Tiahuanaco gain their knowledge thirty centuries ago? To erect such massive monoliths, the Lansburgs assert, would require "a colossal power cranes or some secret of levitation unknown to us."[3] Not only that, but those titanic blocks, some weighing fifty tons and more, were so precisely cut and interlocked that no mortar was needed to bind them. Even today one cannot pound a chisel between them. All this was accomplished without power cranes or hydraulic lifts, elephants or oxen. How on earth, then could they have done it? The only answer possible is, "With nothing we know of on Earth."[4]

Above the Bay of Pisco a giant pictograph in the rock, measuring about 820 feet in size, forms a mammoth trident pointing heavenwards. The pictograph could not have been planned from the ground, for no engineer could have seen his workmen on the rock-face above.

To view it, one has to fly above it. Yet when it was made two millennia ago, men supposedly had no means of flight.

Visitors to the megaliths of Stonehenge have been mystified by a dozen riddles. One observer, Andrew Thomas, asks how could "men wearing skins have designed and erected this computer in stone?"[5]

Greatest Riddle of All

Where did the ancient Egyptians learn the secret of pyramid construction? Sir Flinders Petrie called the Pyramid of Cheops, "the greatest and most accurate structure the world has ever seen." And the Encyclopedia Britannica states that "the brain power to which it testifies is as great as that of any modern man." That is an understatement! No one today could build the Great Pyramid.

This great pyramid is undoubtedly the greatest of the Seven Wonders of the World. Practically everything about it is awesome. The length of each side base is 365.2442 cubits—exactly the number of days in the solar year, including the extra day each four years! The gradient of the pyramid is 10 to 9; for every ten feet that one ascends, one rises in altitude nine feet. Multiply the altitude of the pyramid by 10, raised to the ninth power, and you have 91,840,000 which equals the number of miles from the earth to the sun! This also means that exactly 1,000 pyramids would reach to the sun! And these are not the only amazing measurements found in the Great Pyramid.

Cartographers have discovered that it stands exactly in the land center of the world as it is known today. The Great Pyramid stands midway between the west coast of Mexico and the east coast of China, between North Cape of Norway and Cape of Good Hope of South Africa. It stands at the intersection of the 30th parallel, both latitude and longitude.

So incredible are some of these facts that the most skeptical of scientists have had to admit that the pyramid's location and dimensions can never be a coincidence.

Alan and Sally Lansburg have supplied us with even more incredible statistics.

Somehow its builders knew that the world was round but flattened at the poles, which cause a degree of latitude to lengthen at the top and bottom of the planet; that it rotated in one day on an axis tilted 23.5 degrees to the ecliptic, causing night and day, and that this tilt caused the seasons; that Earth circled the sun once in a year of 365 and a fraction days.

The designers must also have known that Earth's celestial North Pole described a slow circle around the pole of the ecliptic, making the constellations in the sky appear to "slip backward" (the precession of the equinoxes) and bring a new constellation of the zodiac behind the sun at the equinox approximately every twenty-two hundred years in a grand cycle of about twenty-six thousand years. These facts too were part of the internal measurements of the pyramid.[6]

Puzzles Remain Unanswered

Now comes the obvious question: how could the designers of this pyramid have observed the stars for twenty-two hundred years? Another puzzler: how could they have computed, 5000 years ago, the circumference of our planet? And yet another: how could they have measured the rate at which the Earth revolved on its axis? Also, how did they know of the existence of a leap year every four years? Although this information did not become part of our knowledge until six hundred years ago, the pyramid builder knew it more than four thousand years earlier.

One wonders whether the prophet Isaiah had this pyramid in mind when he said,

"In that day shall there be an altar to the Lord in the midst of the land of Egypt, and a pillar at the border thereof to the Lord. And it shall be for a sign and for a witness unto the Lord of Hosts…" (Isaiah 19:19, 20).

Further questions intrigue us: how were those gigantic constructions put together? How were those mammoth blocks of stone, some weighing twenty tons, cut, polished, and made to fit perfectly into place? Even today, the seams between the blocks are hardly visible; and all this accomplished without cement. In all, more than two million blocks of stone, enough to build a six-foot high wall from Los Angeles to New York were used.

The height of the Great Pyramid is equal to that of a 42-story building, and its weight is six and a half million tons. Yet the ground has never given way under it, nor has the pyramid itself moved a quarter of an inch in five thousand years.

Who engineered this mammoth undertaking? And to what purpose? After the endless research and experimentation, we still do not know the answers. In spite of our technology and know-how, our computers and electronic gadgetry, the mystery of the pyramids remains. It is doubtful if modern man could build any of these pyramids, certainly not the Great Pyramid. An attempt was made a few years ago by a Japanese firm but, like others before it, the project ended in failure.

How does one account for this awesome display of knowledge by our pre-historic ancestors? Where does one find the key to such cosmic riddles? If my thesis is correct the key may appear in the near future.

Hollywood Scenarios

It is clear that our generation, more than any in history, has focused its attention on the future. Hollywood, always alert to man's dreams and nightmares, has reflected this truth in a spate of contemporary films. Its two most popular themes during the past two decades have been those of terrestrial disaster and extraterrestrial invasion.

Disaster has become one of the preoccupations of the movie moguls. So much so, that a new word has found its way into our dictionaries—disaster-mania. Historian Christopher Lasch sums it up by saying: "Storm warning, portents, hints of catastrophe, haunt our times."[7]

"Hints" is an understatement. Stories and movies of catastrophe and calamity have become a deluge. Such motion pictures as Earthquake, The Towering Inferno, The Last Wave, Meteor, The Day the World Ended, and The Late Great Planet Earth have followed each other in rapid succession. All were blockbusters at the box office. "Disaster flicks have been the most profitable genre of the 1970s. Nor is the deluge tapering off," reports Time Magazine. These films obviously have some entertainment-value and a lot of curiosity-value, but they also mirror the apprehension rooted deeply in the American psyche.

Many of the disasters portrayed by Hollywood are parallels of calamities predicted in the Bible.

According to Luke 21:9–11, the major signs would include:

Wars

It is claimed that wars have characterized the human family throughout recorded history. There has never been a year without a war somewhere. But our generation has seen an increase in the intensity, frequency, and scope of war never before imagined by the most bloodthirsty tyrant. Dennis Healey, British Defense Minister in the Labour Government, said on TV in 1967: "This has been the most violent century in history. There has not been a single day since the end of World War II when hundreds of people have not been killed by the military action."

Our generation, also, has witnessed the introduction of a completely new element: weapons with the capability of destroying all life on the planet. Science in the twentieth century has made the whole world unsafe to live in. There is no shaded grove or subterranean cave where man can safely hide.

In Matthew 24:6 Jesus warns: "And ye shall hear of wars and rumors of wars: see that ye be not troubled, for all these things must come to pass, but the end is not yet." Commenting on this passage, R. L. Hymers asserts: "It actually predicts war weapons capable of ending the world! The verse makes no sense unless it refers to wars that will make people afraid that the world will end by war. Yet no one had such fear until 1945 and the advent of atomic weapons.

Jesus predicted such wars 2,000 year earlier...[8]

Famines

Skeptics maintain that the famines predicted by Jesus in Matthew 24:7 have been common occurrences throughout history. But again, there is a difference. They have never been on the scale or with the frequency witnessed in the last few generations. *The New York Times* reports that every 8.6 seconds someone dies of malnutrition...10,000 every day...over 3,500,000 every year.

Aggravating the situation is the frightening increase in the world population. It has more than doubled in the last 50 years. Our planet, according to Leighton Ford, has become a "global sardine can."

In many areas of the Third World this situation has been further aggravated by the practices of Eastern religions. In India, for example, Hindus believed in incarnation. That is, many people come back as cows, mice or some other animal. Because of this, cows are sacred and never slaughtered. Every cow eats enough food to feed seven persons—and India has two hundred million cows. The bottom line is that if these animals were slaughtered for food, millions in India would be well fed, and the abundant surplus could be exported to other starving nations. But as long as such beliefs continue, famine will persist as a grim spectre.

Earthquakes

The greatest single disaster known is an earthquake. Like war and famine they have been around a long time, but they also have increased in their intensity and frequency. Dr. Stanley Cronkite, in his historical survey of earthquakes, says "of the thirteen greatest known earthquakes, ten have occurred in the twentieth century." It is not without reason that Jeffrey Goodman called us "the Earthquake Generation."

The United States witnesses a large number of earthquakes each year—California alone averages more than one hundred—yet the states have never really had a truly terrible trembler.

The much-publicized San Francisco quake in 1906 killed seven hundred persons, but the 1976 earthquake of Tangshan, China, killed 655,000.

In addition a new and grim element has been added to the earthquake scenario of our day: manmade quakes have become scientific possibilities. Dr. Cronkite, quoting from an Audio Letter of Dr. Peter Beter, says:

> The matter of China has for sometime been the number one item on the Kremlin prewar agenda. Last month (August 1978) Chairman Hua of China visited Rumania and Yugoslavia in connection with Russian overtures for a reconciliation with China...When Hua met with Rumania's Premier, he was offered certain positive inducements on behalf of the Soviet Union. However, he was also informed that Russia is determined to move fast in its preparation for war against the West. Hua was told that if China is to restore its ties with Russia on a favorable basis, it must be done quickly...Accordingly, Hua was informed that the Soviet Union would shortly provide him with a series of three demonstrations of Russia's operation geophysical warfare arsenal.
>
> The first demonstration was to be an undersea earthquake in the vicinity of Taiwan scheduled for September 2, 1978. The second, to take place the following day, was to be an earthquake on land in the area of Western Europe that is not generally known for earthquake activity. Both of these, Hua was told, would have the intensities in a Richter scale range of 6.5...

> The third demonstration, the finale, was to be an earthquake centered on the Caspian Sea coast of Iran, north of Tehran... About September 14. However, the exact strength...was not predicted...The reason given was that the Iran quake would be brought about by remote means, whose results are not yet highly predictable.
>
> On Saturday, September 2, an undersea earthquake, measuring 6.6 on the Richter scale, occurred near Taiwan, right on schedule...The next morning Southern Germany was rocked by the most powerful earthquake in 35 years, measured about 6 on the Richter scale. Eleven days later, on September 14...a fantastic earthquake, reaching 7.7 on the Richter scale, erupted some 400 miles southeast of Tehran (instead of hitting north of Tehran.)
>
> In moments the town there ceased to exist and the devastation was heavy over a wide area. Within days the death count had reached 26,000 and was still climbing.

In case one should conclude this to be a farfetched, unproven theory, no less an authority than Zbigniew Brezinski (Assistant to President Carter for National Security Affairs) refers to this very possibility in his book, *Between Two Ages.* "Not only have new weapons been developed, but some of the basic concepts of geography and strategy have been fundamentally altered. Space and **weather control** have replaced Suez and Gibraltar as key elements of strategy."[9]

In a similar vein, Dr. Gordon J.F. Mac Donald of the University of California in Los Angeles, in an article in Geophysical Warfare[10] underscores the techniques of earthquake modifications and how they may well be used "in support of national ambitions." Zbigniew Brezinski quotes him as saying, "Accurately timed...Electronic strokes could lead to a pattern of oscillations that produce relatively huge power levels over certain regions of the earth."

Weather control and earthquake manipulation will undoubtedly be some of the grim and devastating weapons of war in the future.

The Bible predicts that the greatest quake of all time is yet to happen. Even the Shensi catastrophe of 1556 and the Tangshan disaster of 1976 will pale into insignificance compared with it. This event is graphically foretold in Revelation 16:18–20. "And there were voices, and thunders, and lightnings; and there was a great earthquake, such as was not since men were upon the earth, so mighty an earthquake, and so great. And the city was divided into three parts, and the city of nations fell: and great Babylon came in remembrance before God, to give unto her the cup of the wine of the fierceness of his wrath. And every island fled away, and the mountains were not found."

The accumulation of disasters in our day finds an increasing number of people echoing the words of President Reagan of the United States: "I turn back to your ancient prophets in the Old Testament and the signs foretelling Armageddon, and I find myself wondering if—if we're the generation that is going to see that come about. I don't know if you've noted any of those prophecies lately, but, believe me, they certainly describe the times we're going through."[11]

These disaster warnings of the end-time predicted in the Bible have received full treatment from Hollywood's special-effects technicians. All signs, that is, except one. There is a solitary sign that has eluded the film experts thus far. It may be the only instance where

Hollywood has not yet caught up with the Bible.

2

The Missing Sign

A century of progressivism bears the fruit of Hitler; our own super-technology breeds witches and warlocks from the loins of engineers.

—John Updike

Hints of this end-time phenomenon have appeared in *The Devil Within Her*, *The Omen*, *The Exorcist*, and especially in *Rosemary's Baby*.

The Exorcist portrayed in harsh, garish color the story of a young lady possessed by a demon. The crowds who flocked to see this film had no difficulty in accepting its credibility; they were completely convinced by it. It portrayed a 12-year-old, innocent girl who played with an Ouija Board and ended up as a degenerate monster, mouthing vulgarities and screeching obscenities. Her face developed into a mass of blood and welts and oozing puss. She engaged in vile and repulsive scenes, and made violent sexual suggestions—even toward her mother.

The story of this demoniac was so realistically portrayed that patrons were terrified in their seats. Some of them even became sick and vomited on the theater floor. Other after-effects suffered by viewers included loss of sleep and appetite, and horrible nightmares. Johnny Carson admitted on his Tonight Show that he did not sleep the night he saw the film.

Paul Scott, the syndicated columnist, called this motion picture the most explosively dangerous film ever produced in the United States. He cited 80 documented cases where persons, after reading the book, themselves became possessed.

Dr. James C. Bozzuto, assistant psychiatrist at the University of Connecticut, warned that these films produce a cinematic neurosis, "significant psychiatric impairment, and both the physician and the public should be aware..." Mary Knoblauch sums it up for all when she says, "Perhaps the most frightening thing about "The Exorcist" is that thirst for and fascination with evil that lies buried within us all, surfacing with savage swiftness at the right incarnation."

Some years before *The Exorcist*, the film *Rosemary's Baby* appeared. The theme of this movie came even closer to the ultimate sign of the coming end. It told of a baby being conceived by a demon father and a human mother. The hint was there, but how many of those who saw the film made the connection with one of the most repulsive signs of end-times?

Science Fiction: Theological Overtones

Another clue to the mystery lies in the world's fascination with UFOs. Early in the 1970s, *Star Wars* and *Close Encounters of the Third Kind* were record breakers at the box office. Spielberg's masterpieces were superb in their scientific gadgetry and realistic effects. Ray Bradbury called *Close Encounters* the greatest film of the last twenty years. It portrayed the arrival upon Earth of alien beings from outer space.

These films and their sequels, aroused millenarian expectations. They were science fiction with theological overtones. Buttressed by a sense of reality, they were no longer considered far-fetched, merely for a good laugh and chuckle. Rather they foreshadowed sober and serious possibilities within our life span.

It had all begun, of course, with science fiction. Edgar Allen Poe wrote about it, with all his creative and analytical genius, using the short story as his literary vehicle. Jules Verne elaborated with his fantastic stories of incredible journeys. Then H. G. Wells came on the scene with his *War of the Worlds.* To many people he was the "guru" of science fiction. "Yet across the gulf of space, minds that are to our minds as ours are to those of the beasts that perish, intellects vast and cool and unsympathetic, regarded this earth with envious eyes, and slowly and surely drew their plans against us. And early in the twentieth century came the great disillusionment" (*War of the Worlds*). This was the book that formed the basis of Orson Wells' radio hoax in 1938, when panic-stricken multitudes rushed out of their homes, convinced that the planet had been invaded.

H. P. Lovecraft was another in a string of science-fiction writers. Some of his ideas were based on biblical statements of the conflict between good and evil. He wrote that the earth had been "inhabited long ago by gross creatures of vastly superior intellects which were presently sleeping but soon to awake and regain the planet."[12]

The Overlords

Arthur C. Clarke was another giant in the field. In his masterpiece, *Childhood's End* (1953), he deals with the last generation of man upon the Earth. It is a generation which sees its offspring transformed into something totally nonhuman and yet superior to humanity. This book became a catalyst for many more such works, developing, and elaborating on a similar theme. According to Woodrow Nichols and Brooke Alexander, "It's probably not too much to say that Clarke's novel became a cornerstone for the developing world view of a whole generation."[13]

It is interesting to note that Clarke added a cryptic announcement in italics on the copyright page of every edition of his book: "The opinions expressed in this book are not those of the author." Among other things he mentions UFO spaceships arriving on planet Earth and crewed by "the Overlords."

These "Overlords" were far superior to man in intellect and proceeded to establish a dictatorship upon Earth. It was orderly and benevolent, but in time the Overlords began to reveal their true character, and lo! "They look like devils." Actually, however, they were supposed to be high-minded guardians of humanity, ruling on behalf of the mysterious "Overmind." They were here to guide mankind through "enormous and inconceivable evolutionary transformations." The new race they produced was the "children of the last humans. At least they look like children, but their faces are devoid of personality, for they are not individuals. They are cells in the body of divinity, neurons in a single mind."[14]

Religion even played a role in Clarke's fantasy. He had an uncanny gift of bringing all the strands together: science, mysticism, religion, parapsychology, and Ufology. The leader of the Overlords brings into focus this significant mind-set:

> Believe me, it gives us no pleasure to destroy men's faiths, but **all** the world's religions cannot be right, and they know it. Sooner or later, man has to learn the truth...Your mystics, although they were lost in their own delusions, had seen part of the truth. There are powers of the mind, and powers behind the mind, which your science could never have brought within its framework without shattering it entirely.
>
> All down the ages there have been countless reports of strange phenomena—poltergeists, telepathy, precognition—which you had named but never explained...But they exist, and if it is to be complete, any theory of the universe must account for them.

It is not difficult to note the underlying aim of science fiction. According to the assessment of John Keel,[15] it has served to "program" and "condition" mankind for the real thing.

Science Fiction to Science Fact

In our day, the real thing has unquestionably arrived. Our generation has taken the leap – and a giant one at that—from science fiction to science fact. An age which has witnessed man stepping on the moon and start on his star-trek to the distant galaxies, finds it not too difficult to believe in a reverse trek from distant worlds to planet Earth.

No subject has greater fascination for modern man than Ufology, and no subject has spawned more theories as to the origin and identity of the space vehicles, commonly called UFOs, which appear to be invading our world. A strange phenomenon is recorded in Genesis 6, which could prove to be the missing clue in solving the UFO mystery.

> When men began to increase in number on the earth and daughters were born to them, the sons, of God saw that the daughters of men were beautiful; and they married any of them they chose. The Nephilim were on the earth in those days—and also afterward—when the sons of God went to the daughters of men, and had children by them. They were the heroes of old, men of renown.
>
> Genesis 6:1, 2, 4. (NIV)

Controversy surrounds the interpretation of this passage. Nevertheless, strong biblical evidence indicates that it refers to the bizarre union between extraterrestrials and the women of Earth. The Nephilim were the superhuman offspring of the union, and they appeared on this planet just before the great Flood. In fact, their existence and vile corruption of the world was the main reason for the catastrophe. Their kind was destroyed along with the rest of mankind in the Flood. Only Noah and his family escaped their contamination and hence were saved.

Yet, centuries later, the Nephilim emerged again, this time on a more limited scale in the land of Canaan (Numbers 13:2, 25–33). As before, God ordered their destruction.

Threat of the Nephilim

This immediately brings to mind a number of intriguing questions. Why is the presence of the Nephilim so great a threat to man that God would resort to such drastic measures? Who were the parents of these superhumans, and where did they go? The reappearance of their progeny would indicate that they returned to again contaminate the human race with their offspring. Evidence indicates that some of these Nephilim survived the second extermination. If so, where are they today? And are their extraterrestrial parents the aliens responsible for the UFOs?

If the procreators of the Nephilim are to reinfest our world successfully, they must prepare mankind for the assault, for they face a more sophisticated human race. What greater ploy than to convince us that they are brothers of man travelling from distant worlds to invite us into the great galactic society? What greater stratagem than to awaken our imagination to the possibility of a supernatural union between aliens and the "daughters of man."

Rosemary's Baby and *The Devil Within Her* are but one facet of the deception. Hollywood may soon make the connection between this subtle sign of end-times and the apparent arrival of beings from another dimension.

It is significant that, after a long struggle for recognition, millions of people now believe that UFOs exist, and what is more that they have been on our planet.

These space vehicles have been tracked on radar, fired on by jet fighter planes, photographed a hundred times, and have left indisputable evidence of their landing on the ground. In all, fifteen million Americans, including a former President of the United States, claim to have seen them. The evidence is such that Marcia Seligson could write in the *New West* magazine that "there is an accumulation of material weighty enough to bury the toughest skeptic."[16]

If one can accept the incredible idea that "brothers of man" are indeed making contact with us from outer space, mankind is one step closer to a frightful discovery.

3

Close Encounters

If the government has been
covering up Watergate, then their handling
of UFOs is a cosmic Watergate.

—J. Allen Hynek

While multitudes are convinced that we are being visited by cosmic travelers, millions of others view UFOs as nothing more than a hoax. To them, the evidence offered by people around the world is too incredible to believe. Even on the official level, doubt prevails. Despite this skepticism, the UFO controversy has proven one thing—neither the state nor the science can explain the phenomenon. Unwilling to admit their failure, many have resorted to the simple expediency of denying that UFOs exist.

Many of the scientists on the White Sands Missile Range in New Mexico would no doubt disclaim any belief in Ufology. But words found there scribbled by an unknown hand on a restroom wall may well be nearer to the core of truth:

I saw a disk up in the air
A silver disk that wasn't there.

Two more weren't there again today
Oh I wish they'd go away.[17]

Finding themselves unable to deny the phenomenon completely, other scientists have resorted to various rational or semirational arguments. Some assert that witnesses of UFOs are mistaken; others have explained them as optical illusions, plasma discharges, swamp gas, or the reflection of the sun through thin ice fog.

Many opt for a simpler solution. UFOs are nothing other than secret, military spacecraft, designed and operated by one of the superpowers. If one could get close enough to one of these "flying saucers" and be able to read the markings, he would no doubt see, "Made in the U.S.A." or "Made in the USSR." This too, would explain the official government silence on the subject. It is nothing other than "Washington's cosmic cover-up" according to one author.

Major Murphy, a former high official in the U.S. Intelligence, suggests that UFOs are not spacecraft at all but rather "psychotronic devices."[18] He holds that during World War II, the Germans were doing advanced research on controlled electrical discharges and "controlled lightning" and tried to combine the two. He also claims that the scientists were working on "circular aircraft" and were hoping to develop a new and secret weapon. In an interview with Jaques Vallee, Major Murphy said, "When we invaded Germany, a lot of hardware fell into our hands, but the Russians had gotten most of the good stuff. It was soon after this that people started seeing UFOs, 1946 in Sweden and 1947 in America."[19]

Are UFOs for Real?

All suggestions and theories to the contrary, the reality of UFOs is becoming increasingly obvious. No theory has been able to dispense with the evidence for their existence. The space vehicles simply will not go away.

Dr. J. Allen Hynek, chairman of the Astronomy Department of North Western University, tells us that "there is sufficient evidence to suggest we are not alone." What adds to Dr. Hynek's credibility is that he was at one time a noted skeptic himself! But now, after intensive and exhaustive research, he concludes that most UFO reports are from "reliable, excellent witnesses." "When I first got involved in this field, I was particularly skeptical of people who said they had seen UFOs on several occasions and totally incredulous about those who claimed to have been taken aboard one. But I've had to change my mind."[20]

Others have experienced a similar change of mind. When the prestigious American Astronomical Society conducted a survey of its members, an impressive 53 percent said that UFOs "certainly" or "probably" should be investigated further. When a sample poll of U.S. Citizens was taken in the 50s, only 3.4 percent believed in UFOs and that they were manned by extraterrestrial beings. By 1978 that percentage had risen to 57 percent.

What happened was undoubtedly this: millions began with outright disbelief, then as the data increased and more evidence became available, the disbelief dissolved into doubt. Later, as the data kept increasing, and the evidence simply would not go away, the doubt turned to wonder. That seems to be the stage that multitudes have reached today; unable on one hand to fully explain the evidence, and unable on the other hand to explain it away, they are left suspended in a state of wonderment.

Their situation is not unlike that described by Professor Huston Smith of Syracuse University. "The larger the island of knowledge, the longer the shoreline of wonder."[21]

What Is the Evidence?

What proof do we have that such an "island of knowledge" exists? What is this increasing evidence for the UFO phenomenon?

First, the sheer weight of the facts involved. No less than fifteen million Americans—11 percent of the adult population—claim to have seen a UFO. Among them was a former President of the United

States, Jimmy Carter. He filed two reports of UFO sightings when he was Governor of Georgia. Other notable men who were convinced that UFOs needed serious consideration were General Douglas MacArthur, President Dwight D. Eisenhower, Senator Barry Goldwater, Lester B. Pearson (former Prime Minister of Canada), Malcom Muggeridge of England, and Thor Heyerdahl of Kon-Tiki fame. Considering the numbers involved, one should remember that 90 percent of the people who claim to have seen a UFO never report the fact. They belong to what Dr. Hynek calls the "Legion of the Bewildered Silent." However, those who do report it average some one hundred sightings every twenty-four hours!

Dr, Hynek asserts that he now possesses a computer-bank of 63,000 sightings.[22] The Australian computer-bank has passed the 80,000 mark. And that was in 1977! And even this may well be small compared with the files of Aimé Michel, whose files on UFO sightings were so extensive that it took four years to analyze them in any detail!

It should be added that many of these UFO sightings were viewed by dozens, hundreds, even thousands of people at the same time. During the numerous UFO landings in France in 1954, only 15 percent of the witnesses were alone at the time; all the others were multiple-witness cases. It is also significant that independent evidence given by various groups which had never heard of each other, agreed on a high percentage of facts and details. The truth is, according to Dr. Stanton T. Friedman, nuclear physicist and UFO researcher, that there is a "Close Encounter of the Third Kind" occurring somewhere on this planet every day. This is in addition to "Close Encounters of the First and Second Kind." It would be strange indeed if all this evidence were the result of mass hysteria. Ian McLennan makes it clear when he says: "Hundreds of thousands of people cannot make independent observations of the same optical phenomenon over thousands of square miles under a fantastic dose of mass hysteria."[23]

A typical example of a multiple-witness sighting would be the one over Indianapolis, July 13, 1952. A huge, oval-shaped space vehicle was seen racing over the city about 5,000 feet above the ground and witnessed by thousands of spectators at the same time. Suddenly the switchboards at police stations, newspaper offices and the airport were swamped with frightened inquiries. If further corroboration were needed, Air Force radar had picked up a high-flying craft in the area, just before the actual sighting.24 And for good measure, pilots from American Airlines, Eastern Airlines, and the U.S. Air Force testified to seeing this same object traveling at a fantastic speed.

Are the Witnesses Reliable?

Are the witnesses of UFOs reliable? Can they be trusted? Would their testimony hold up in court?

The variety of witnesses is endless. They include civilians from every strata and society, not crackpots or lunatics but rather normal men, family men, many of whom hold responsible positions in their communities. There are also numberless military witnesses from the Air Force, Navy and Marine Corps—radar men, guided missile trackers and ground troops. Aeronautical engineers, airport traffic controllers, weather bureau observers, ground observer corps members, astronomers, FBI agents, state, county and city police, official photographers, plus private pilots and crews from American, United, Eastern, Pan American, Northwest, Western, and Trans World Airlines are also on the list of witnesses. In the words of Dr. Clifford Wilson, "To an unprejudiced investigator it would have been hard to find a group better qualified to observe and report on UFOs."[25]

It is of significance that the same kind of descriptions are being given both by trained U.S. airline pilots and by half-civilized, remote islanders from the far corners of the earth. Some of the witnesses had never heard of a flying-saucer. An example would be Celeste Simonutti, and Italian working on a tiny island off the French Coast, the Isle of Re. As he was returning home on the night of September

30, 1954, he saw a brilliant light. Thinking it was a fire, he hurried to the scene. What he saw was a luminous sphere about twelve meters in diameter floating less than one meter above the ground. The sphere became red, then blue, then took off vertically at extreme high speed. [26] The interesting thing is this: when people told Celeste Simonutti he had seen a flying-saucer, they had to explain to him what flying saucers were—he had never heard of them! He was an Italian, living in France, spoke very little French—and never read a newspaper!

One should also add that some of the witnesses were press reporters, a category of people who would traditionally be skeptical. Another class that could be described as traditionally skeptical would be scientists, interested only in hard-core data. Today a number of these scientists have had to admit that they had to "suspend their innate skepticism." One such skeptic, as already noted, is Dr. J. Allen Hynek. He was the official astronomical consultant to "Project Blue Book"— the U.S. instituted enquiry into the existence of UFOs. Dr. Hynek, a scientist of impeccable repute, headed the NASA-sponsored satellite-tracking program at the Smithsonian Astrophysical Observatory in Cambridge, Massachusetts. In all, Dr. Hynek has spent more than twenty years studying the subject of UFOs. He describes his early skepticism: "Before I began my association with the Air Force, I had joined my scientific colleagues in many a hearty guffaw at the 'psychological postwar craze' for flying saucers that seemed to be sweeping the country and at the naivete and gullibility of our fellow human beings who were being taken in by such obvious nonsense."[27]

To Dr. Hynek, at that time, all talk about UFOs was "nonsense" and "nonscience." But the same Dr. Hynek, twenty years later adopts a far different stance: "It is no longer possible to sweep away the whole subject. It reminds me of the days of Galileo when he was trying to get people to look at the sun spots. They would say that the sun is a symbol of God; God is perfect; therefore the sun is perfect; therefore spots cannot exist: therefore there is no point in looking."[28] Today, he states quite categorically: "I can establish beyond a reasonable doubt that they (UFOs) are not **all** misperceptions of hoaxes."[29] Another leading scientific specialist at the University of California recently stated: "I know of no scientist who has become professionally involved with UFO investigation who doesn't believe in the extraterrestrial phenomena."

International Complement

Witnesses are not confined to America. Millions in the Soviet Union, the United Kingdom, France, Australia, South America, Mexico, and nations around the world have seen UFOs. Jaques Vallee has in his files cases of Close Encounters *from every country on Earth*, many of them involving UFO occupants of various sizes and shapes.

Astronaut Gordon Cooper claims that he saw hundreds of UFOs in Europe and chased them in his plane for hours.

According to Dr. Olavo T. Fontes, professor of medicine at the Brazil National School of Medicine, there is no UFO controversy in Brazil. Such vast numbers have seen UFOs at close range that they need no convincing. Dr. Fontes personally investigated more than three hundred of these cases.

Robert Gallery, Minister of Defense for the French Government, said on the radio: "If your listeners could see for themselves the mass of reports (concerning landing UFOs with occupants) coming in from the airborne gendarmerie...They would see that it is all pretty disturbing. In fact, the number of these gendarmerie reports is very great."[30] And Jaques Vallee lists 923 documented cases of UFO landings in France between 1868 and 1968.

Other accounts of UFO sightings have been reported from Russia, and Soviet Deputy Premier Kozlov is one of the men there who is convinced of their existence. According to Brad Steiger and Joan Whriteow in their book Flying Saucers are Hostile: "In the Spring of 1959, UFOs brought more panic to Soviet Radar and Air Force personnel by hovering and circling for more than 24 hours above

Sverdlovsk, headquarters of a tactical missile command. The Red fighter pilot sent aloft to chase the UFOs away reported that the alien objects easily outmaneuvered their jets and zigzagged to avoid their machine gun fire. Dozens of nervous candidates for Soviet civilian flying licenses have complained about UFOs sweeping at them and even following their planes back to the airfields."[31]

There have been so many sightings in the Soviet Union that they too have launched an official UFO study project under the leadership of Major General Porofiry A. Stolyarov. When the two Russian cosmonauts, Uri Romanenko and Gueorgui Crechki, returned to Earth in March 1978, after ninety-six days in orbit aboard their Salyut 6 orbiting spacecraft, they told of a formation of UFOs which trailed them closely for three complete orbits around the globe. One of the astronaut's cameras filmed a twenty-minute fantastic motion picture footage of this encounter. A second camera took still shots, securing some outstanding pictures. This news, of course, was suppressed in the Soviet Union, but they did—according to a previous mutual agreement—forward copies to the United States to be analyzed by NASA. Unofficial statements from NASA state that it is the best motion picture footage ever filmed of UFOs. And some of the still photographs have been enlarged to four feet square, revealing many features of UFOs. A report leaking out of Russia states that the whole episode "marks irrefutable proof of their (UFOs) existence and of their extremely high maneuverability in space."[32]

Paul Vorohaef, a Russian not living in the United States, says that Russian peasants interpret the "saucers" as awesome warnings from God and as a sign of the end of the world.

Dates and Data

UFO sightings cover a vast period of time. For thousands of years there have been reports of strange sightings and records of such sightings have been found in a number of ancient manuscripts. One of the earliest, undoubtedly, was found in Egypt, dated about 1,500 BC. It referred to a "great circle of fire…Coming in the sky" and followed by other similar circles. These vehicles had no head and no tail, but they did have a mouth—but a mouth with no voice! As good a description as any of the spherical craft.

The modern furor concerning spacecrafts seems to have begun in 1947. It was in that year that Kenneth Arnold was flying in his private plane from Chehalis to Yakima, Washington when he sighted nine gleaming discs near the Cascade Mountains. The nine were strung one behind the other traveling at fantastic speeds, their estimated size about one hundred feet in diameter, and "saucer-like" in shape. The name caught on and soon "flying-saucers" became a household term in America. Arnold's name found its way into our history books. Soon, reports of other sightings began coming in from different parts of the country confirming his experience.

Such sightings, however, had been recorded much earlier then 1947. As far back as 1896–7 there were records of UFOs appearing over San Francisco and Chicago. More than 150 sightings were reported during that time. The records refer to space vehicles as "bizarre, lighter-than-air ships…Sporting propellers, port-hole windows, antennae and brilliant searchlights which they directed at the ground."[33]

On May 15, 1879, a strange entry was made in the log of the HMS Vulture when cruising in the Persian Gulf. Commander Pringle recorded that two incredible objects, glowing brightly, were pacing his ship, one on each side. He described them as revolving wheels with illuminated spokes.[34]

Long before then, American Indian mythology had described space vehicles in its own characteristic fashion. The Sioux Indians referred to "sky people returning to their home on high by turning themselves into arrows and ascending in that form."[35] Some men have used this description as proof of extraterrestrial visitors being able to adapt themselves to the cultural milieu of the time.

For thousands of years there have been reports of sightings of strange vehicles in the sky, and these records have been preserved in ancient manuscripts. One author, Ed Vallowe on a tape message says every generation in history, except one, has a recorded reference to a UFO—the one exception being the generation when Christ walked on this Earth, AD 0–30. This is a highly significant fact in view of the thesis we are proposing. It seems too easy to dismiss all these reports as mere legends. Persistent legend, duplicated in different countries and cultures, would seem to indicate a basis of fact. Serious space writers are today giving much attention to these reports, and are finding certain similarities between primitive and present day descriptions of UFO sightings.

The proliferation of investigative organizations, both official and private, also would indicate a basis of fact to UFO reports. Whereas the governments of many countries have set up organizations to study these sightings, far more numerous are the private organizations established for the same purpose. Typical of these organizations would be the Ground Saucer Watch of Phoenix, Arizona and the Project Starlight International of Austin, Texas. One hundred such organizations exist in America alone.

Such is the interest in the subject that a new science has come into being, called "exobiology." Its function is to search the universe for extraterrestrial life. Its leading spokesman is the noted Carl Sagan of Cornell University. His books, *The Cosmic Connection* and *The Dragons of Eden* have become best sellers. Sagan is convinced that civilizations exist in the reaches of space. He produced a remarkable LP record of "earth sounds" that was fastened to the space ship Voyager for its trip into space, far beyond our solar system. In the record Sagan included greetings in sixty languages, including sounds from volcanoes, avalanches, surf, and animals. There is even the sounds of whales saying, "Hi there!"

More Dates and Data

The existence of photographs and certain physical records are further proof that UFOs exist. Rex Heflin took photographs of a UFO in Orange County, California, in 1965, when he was a highway traffic investigator for the County Road Department. He saw this vehicle skim over telephone poles, cross the road in front of him and continue its flight on a low level. The sighting lasted about fifteen seconds during which time he took four Polaroid photographs. These photographs were later published by the *Santa Ana Register* and released to other papers by United Press International.[36]

No less an expert than Arthur Lundahl, former director of the Photographic Interpretation Center of the CIA, has issued a statement that he had examined a number of UFO films on behalf of the U.S. government and found them to be authentic.

Astronaut John McDivitt took movies of a UFO as he orbited the Earth in 1965 at a speed of 17,000 miles per hour. The strange space-vehicle, disk-like in shape, circled their craft as he and Astronaut White were on their twentieth revolution. He has no doubt what ever as to the craft he saw and filmed.

In addition to this, photographs exist of physical marks left on the ground by space vehicles. Sometimes these marks are scorched earth, sometimes tripod impressions.

Physicist James McCampbell, addressing a UFO symposium in 1975, said: "Evidence left at landing sights leaves little room that UFOs are heavy, ponderous objects when addressed, yet in flight their startling departures, sudden stops, and right angle turns at high speed require them to be virtually massless."[37] With such evidence one is entitled to ask: How much more is needed before everyone believes in the reality of this phenomenon? One can understand men like Edward J. Reppelt, head of Project Blue Book, being perplexed by the unwillingness of men to accept the evidence. He asks: "What constitutes proof? Does a UFO have to land at the river entrance to the

Pentagon, near the Joint-Chief-of-Staff offices? Or is it proof when a ground radar station detects a UFO, sends a jet to intercept it, the jet pilot sees it, and locks on with his radar…Is it proof when a jet pilot fires at a UFO and sticks to his story even under the threat of court-martial?"[38]

Confronted with the mass of evidence that UFOs really exist, the question that inevitably follows is: How is one to explain it all?

Millions of people identify with the conclusion of Carl G. Jung, the famed founder of the modern psychiatry: "By all human standards it hardly seems possible to doubt (UFO existence) any longer."[39]

Close Encounters

The science of Ufology specifies three distinct categories of sightings:

1. *Close Encounters of the First Kind.* This refers to mere sighting of an unidentified flying object at close quarters.
2. *Close Encounters of the Second Kind.* This includes not only the sighting of a UFO at close quarters but also the presence of certain physical records of the craft having been there. These records may be tangible marks on the ground, the scorching of grass, interference with electric circuits, or physical effects on animals or humans, like temporary paralysis and weightlessness.
3. *Close Encounters of the Third Kind.* This is the most bizarre and "incredible" of the three encounters, for it includes, in addition to the two distinctives above, a direct confrontation with a space-being, or entity, sometimes called a "humanoid." These sightings are so "incredible" that scientists of the caliber of Dr. Hynek would gladly omit them altogether if they could do so, "without offense to scientific intelligence."[40]

There is a tendency on the part of all intelligent beings not to accept evidence of such mind-boggling nature. It means that one has to go out on a limb if he is to subscribe to a belief that animated, living creatures have been seen descending from strange space vehicles gesturing to humans and even trying to engage them in conversation. People have become so accustomed to viewing such encounters as jokes that it is now well-nigh impossible to treat them as facts. After all, who wants to go on record as one that has actually seen a giant humanoid, and had communications with him? Or of having seen little green men peeping through your kitchen window? It is like believing in leprechauns! Such things may be all right for crackpots, but not for men of intelligence. No wonder Aimé Michel referred to this type of UFO phenomenon as a "festival of absurdity."

But in spite of such reaction, Close Encounters of the Third Kind does have one thing in their favor. One may be excused to mistaking or misidentifying an object at a distance, but to do so when it is within a few yards of the observer, is most unlikely. And this is often the case with Close Encounters of the Third Kind.

Sturno, Italy 1977

One of the most recent reports of one of these encounters comes from the little town of Sturno near Naples, Italy. On August 29, 1977, no fewer than seven men from Sturno saw a spaceship parked on the ground and also a giant, space-suited humanoid trying to communicate with them. The men were awestruck and spellbound as this "creature" gestured to them.

Two of the seven witnesses were put into trances and questioned by the medical hypnotists. According to the report of Professor Franco Granone of the University of Turin: "They are not telling lies. The subjects, under hypnosis, have referred to facts they actually lived through."

The space giant was first spotted by two college students, Michele Giovanniello and Rocco Cerullo just after midnight. Here is a part of a newspaper report February 21, 1978. According to Giovanniello:

"We were out walking. Suddenly, at the crossroads of a mule track leading to an abandoned quarry, we were attracted by a glowing red light behind the trees on a hill. We approached it cautiously and we heard strange noises like those in a science fiction film: 'bleep, bleep.' Words cannot describe how I felt! Then we saw the outline of a tall figure. It stepped forward and was lit by the moon's rays. It was like an athlete in a luminous suit which clung to the body, and it had on a silver helmet. We were 35 feet away—petrified by what we saw. Then the figure turned to face us—and from the level of its eyes came flashing lights, red and orange. We turned back to the village, scared stiff!"[41]

Giovanniello and Cerullo excitedly told their story to three men they found in the village square—medical student Antonio Pascucci, school teacher Amalio Capobianco and machinist Arturo D'Ambrosio.

"Our first reaction was to laugh," recalled Pascucci, 24, who attends Naples University. However out of curiosity they went to the scene in D'Ambrosio's car. "The creature was dressed in a silver suit with a belt…Where two eyes should be were two orange lights which blinked…. It had arms and legs, but we could not see if it had normal feet and hands. In the distance, down the slope, we could see the outline of an object with portholes and white lights inside. There was a dome on top with another light."

Pascucci came to the conclusion that the figure defied the laws of gravity as it moved up and down the slope. Its body was always at the right angles to the ground, whereas humans must lean on steep slopes to keep from falling. Pascucci tried to walk in the same way as the figure but fell over each time. He continued: "With its 'eyes' it blinked out messages in Morse-like code. We couldn't understand the messages, but the rhythm was a regular pattern as if it was trying to communicate. Then it tried with high-pitched sounds, like radio—again in Morse form—before changing to a low pitch like an airplane engine."

The five drove back to Sturno, where they came across singer

Mario Sisto and his friend Michelino Riefoli, also out for a late-night stroll. Sisto, 39, and Riefoli, 49, returned to the quarry site with the men and the giant figure was still there. Sisto said that the humanoid, which had a black box attached to his waist, made a beckoning gesture with its right arm "like an invitation to follow." But Sisto and his friends were too afraid to move, so the humanoid began to move toward them. Then it stopped, pointed towards the sky, and seemed to indicate that he came from up there. To find out whether it was a trick, the men threatened it, saying they had a gun and would shoot if he didn't speak up. But the humanoid made no response.

Although scared at the time, the fascination of it all kept the men from running away. Pascucci said that he tried to communicate with the creature using sign language but had no success. The figure came no closer than fifty feet to them.

The men watched the humanoid until about 3:30 a.m. "Then," said Pascucci, "suddenly a ray of intense white light shot out from its forehead, lighting up the whole area."

This time the men were panic-stricken, fearing that the light was a laser beam, and that it would shrivel them! The seven fled back to Sturno. The next morning, villagers flocked up to the site, but by this time the mysterious creature had disappeared. The unidentified craft had also vanished, but in the hard, rocky ground where it had been sighted, three circular imprints arranged in a triangle were visible.

These strange imprints were examined by construction engineer Carmine Cangero, but he was unable to offer any explanation for them.

"Two sides of the triangle measured about 13 feet, and the base measured about 14 feet," he told the newspaper. "The diameter of the circles was about 8 inches. Three holes inside the circles, also in the form of a triangle, were 1 ½ inches deep." By his calculations, he said, it would take an immense force to create the imprints. "The only thing I know of which could possibly have made them is a pile driver, but it would be impossible to get one up the mule track."

All seven men were questioned carefully by the police, by a doctor, and by the mayor of Sturno—and all were convinced they were telling the truth.

Sturno, Italy, 1977, seems a long way off, geographically and chronologically from the events of Genesis 6. But the answer we are seeking is whether there is a linkage between the two. Does the giant seen by men of Sturno and other Encounters of the Third Kind have any relation to the giant Nephilim of pre-Deluge times?

4

Physical or Psychic?

Angel beings shuttle in a flash
from the capital glory of heaven
to earth and back again.

—Billy Graham

Of all the data gathered under the category of Ufology, none seems more absurd than Close Encounters of the Third Kind. To imagine living entities or humanoids riding around in space vehicles and engaging in conversation with humans of Earth is too bizarre to merit serious consideration. Nevertheless, the evidence is there. "The fact is," says Dr. Hynek, "that the occupant encounters cannot be disregarded; they are too numerous."[42] He elaborates: "Despite the bizarre nature of the reports and the seeming impossibility of their having happened, the fundamental question is, as before, not **could** these reported things have happened, but **did** they happen, more or less as reported? I would not be engaged in delineating these matters in this book had not the evidence I personally have examined over the past years seemed overwhelmingly to indicate yes as the answer to the latter question. The bizarre events actually did occur, as unthinkable as this may seem to the physical scientist."[43]

And as for documentation, Dr. Hynek states: "It will undoubtedly surprise most readers to learn that there exists a catalog of some eleven hundred cases in which a UFO occupant has been reported."

These eleven hundred cases are additional, of course, to the scores of thousands of others belonging to the categories of Close Encounters of the First and Second kind.

Jaques Vallee has prepared his own catalogue of 1,247 close encounter cases and found 750 of them referred to an actual landing of an alien craft. And of these, more than three hundred had humanoids visible from the craft, or in close proximity of them. For good measure, he reports that many of these were multiple-witness cases.

HUMANOIDS?

Who are these humanoids? Whence do they come? Whither do they go? What are they doing here? Where is their "home"? Could they be the procreators of a latter day Nephilim?

Christians are divided on the issue. Many, if not most, would say that nothing comparing to human-life can be found on any other planet. Man was made for planet Earth only, and Psalm 115:16 is quoted as confirmation. "The heaven, even the heavens, are the Lord's: but the earth hath he given to the children of men."

Others contest this opinion, and hold that God could have created similar beings on innumerable planets. Why should He keep empty houses?

At the turn of the century, Alice Meynell expressed the belief of many such Christians when she wrote:

No planet knows that this
Our wayside planet, carrying land and wave,
Love and life multiplied and pain and bliss,
Bears, as chief treasure, one forsaken grave…

But, in eternities
Doubtless we shall compare together, hear

A million alien Gospels, in what guise
He trod the Pleiades, the Lyre, the Bear.

Oh, be prepared, my soul!
To read the inconceivable, to scan
The million forms of God those stars unroll
When in our turn, we show to them a Man.[44]

Like Christians, Ufologists are divided on this question. Many believe extraterrestrial beings from other planets visit our planet from time to time. To them, there is no other viable alternative. Unable to dismiss all UFO sightings as hoaxes, hallucinations, or weather phenomena, they have had to come up with a theory. So they have concluded that these "beings" have to be extraterrestrial visitors from distant galaxies. And this being so, they are obviously way ahead of humans in their technology, able to travel by use of electromagnetism or some other unknown means.

Of course, small variations exist in these theories. Some claim that all extraterrestrials come from the same planet. Others believe they come from a number of planets, or even galaxies. Some add that they may come from an entirely different period in that time; that the beings are from the future sent back in time to observe their own beginning! Others have suggested that the extraterrestrials not only come from different galaxies and time-periods, but that their method of travel and communication suggests knowledge of a fourth dimension, or maybe a fifth and sixth dimension, way beyond our comprehension. Still others claim that these beings are not so much extraterrestrials or metaterrestrial; they come from a parallel world to our own.

Cosmic Provincialism

The proponents of these theories suggest that our hesitancy in accepting their viewpoint is to our "cosmic provincialism" and to our "stellar ignorance." Fred Hoyle in his book, *Of Men and Galaxies,*[45] says that it is possible that a great intergalactic communications network exists but that we are like a settler in the wilderness who as yet has no telephone.

Dr. Allen Hynek elaborates on this aspect of our ignorance:

> Let us imagine for a moment a covered wagon train not much more than a century ago. Winding its long journey to the West. It is encamped for the night, its wagons in a circle, sentries posted, and the travelers gathered about a campfire for warmth and cheer. Someone speaks of the future but he speaks, as he must, with the words and concepts of his day. But even were he inspired by some kindly muse of the future to speak of making their entire journey in a matter of house—flying through the air, and of watching scenes by television, and hearing voices speaking on another continent, this gifted one could not have put into words a glimmer of how these wondrous things could be accomplished. The vocabulary for such descriptions—electrons; transistors; integrated circuits; jet engines—the jargon vehicle of technical communications would not yet exist for another century. He would be helplessly incoherent for want of words as vehicles for his thoughts.

Would one care to venture a guess at the technological vocabulary of the year 373,475 (assuming intelligent life still exists on the earth)…?[46]

Think back a mere fifty years and consider the changes that have occurred in that short time. Half a century ago there were no such things as television, penicillin, nylons, microwave ovens, missiles, rockets, even ball-point pens! Similarly, it is claimed, at some future point in time, knowledge of these space-beings will be as commonplace as the technical gadgetry we take for granted now. Today, we merely know that these space-beings exist; we do not know who they are, whence they come or whither they go. That truth has not yet come to us, or we have not yet come to it. Such truth, of course, may take a long time to come; on the other hand, it may dawn suddenly and unexpectedly.

Undiscovered Ocean of Truth

It is increasingly obvious, that a vast reservoir of knowledge still lies beyond our reach, and which continues to remain untapped. The paradox is, that although our knowledge has doubled and trebled in the last few decades, so has the awareness of our ignorance. It was one of the greatest scientists of all time who said: "I do not know what I may appear to the world; but to myself I seem to have been only like a boy playing on the seashore and diverting myself now and then finding a smoother pebble or a prettier shell than ordinary whilst the great ocean of truth shall lay undiscovered before me" (Sir Isaac Newton).

However, none of these theories seem to carry a positive degree of credibility; not with the public nor with men of science. Few scientists, in fact, will accept the hypothesis that UFOs and their occupants are extraterrestrial beings. Dr. Donald H. Menzel and Dr. Ernest H. Taves,[47] while sympathetic to the hypothesis that life may exist on other planets, nevertheless conclude that this is no proof that our Earth is being visited by extraterrestrial spacecraft. In their opinion "the tremendous distance involved means that interstellar travel would be enormously difficult for any would-be space travelers no matter how advanced." We are thus left with the question: Is there another viable, credible alternative?

I believe there is. However, before I introduce that alternative, a number of facts should be noted:

- Not one of these space vehicles or space beings has ever been captured.

- Not one space vehicle has ever crashed on earth. Not one single, tangible piece of evidence—nut, bolt, chunk of a saucer, extra-terrestrial debris, trash can, has ever been found.

  These two facts are seriously denied by a number of experts in the field. They state that a number of space vehicles have been recovered and that many of them are now stored in a secret air base in Nevada.
- Not one has ever responded to radio or any other kind of signal.
- No planet has ever been positively identified as the home-base of UFOs. All the messages seemingly received thus far have indicated many planets, without consistency or any proof of any kind. About the only sensible message given was the one back in 1897, by the occupants of an airship-shaped UFO: "We're from **ANYWHERE**, but we shall be in Cuba tomorrow!"
- No one has explained with any degree of credibility how space vehicles can travel the vast distances involved. If they come from another galaxy, then they come from a locale at least two million light-years away! It boggles the imagination to conceive such a thing taking place. Even traveling close to the speed of light, 186,000 miles per second, it would take a minimum of two million years to get here!

  But supposing they come from a planet within our own galaxy (The Milky Way), it would still take them anywhere from 4–38 light-years to get here. One thing is crystal clear, you can't make that journey by physics—or by any known principle of aerodynamics.

To convey the immensity of interstellar distances in a different way, Walter Sullivan supplies a useful illustration. If we scaled down the sun to the size of a cherry stone, planet Earth would be a grain of sand three feet away. But the nearest star would be another cherry-stone 140 miles away![48]

What is the answer to all this? There is none, unless of course, the extraterrestrials have an entirely different concept of space-time. Some theorists believe this to be so. On the basis of certain Einsteinian propositions, they have suggested that a journey from outer space would be much shorter to extraterrestrial travelers, because time would be decelerated for them. Inside a space vehicle traveling at great speed, a trip that would take decades in "planet-time" would only take days in "cosmonaut time." However, when the extraterrestrial would return home, a few weeks later, he would find his friends and family whom he had left behind, to have been dead a long, long time.

Space Not a Vacuum

But in addition to the problem of space travel, there is another complication. Space is not a vacuum but rather contains cosmic gas, dust, and particles. It is calculated that there is about one hydrogen atom per cubic centimeter, plus bits of dust and other material. A space vehicle traveling at a speed close to that of light would collide with this space-material in the form of radiation—and that with an impact of several hundred atom smashers, says Professor Edward Purcell, the Nobel laureate of Harvard. This, plus the fact that physical beings, no matter how advanced, are still governed by physical laws, makes the very idea preposterous. No wonder Dr. Purcell concludes: "All this stuff about traveling around the universe in space suits—except for **local** exploration (within the solar system)...Belongs back where it came from, on the cereal box."[49]

Assuming that all these obstacles can some day be overcome, we are still back in square one. As G. G. Simpson reminds us: "Not even one earthlike planet outside our solar system has been observed objectively or known to exist in fact. Exobiology is still a "science" without any data and therefore no science."[50]

But supposing that these space-beings by means unknown to us, succeed to arrive here from some distant galaxy and survive the incredible journey, it still has to be explained how they are able to accommodate themselves to the atmosphere and environment of our little planet. How are they able to "blithely enter our atmosphere, breathe our air, speak our languages, and even wear clothes that fit in with the particular cultural patterns where they have landed."[51]

Two of the most respected and qualified researchers on the subject, Vallee and Hynek have completely discarded the theory that they come from another galaxy. "If UFOs are indeed somebody else's nuts and bolts hardware, then we must still explain how such tangible hardware can change shape before our eyes, vanish in a Cheshire can manner,(not even leaving a grin), seemingly melt away in front of us, or apparently "materialize" mysteriously before us without apparent detection by persons nearby or in neighboring towns. We must wonder too, where UFOs are "hiding" when not manifesting themselves to human eyes."[52]

Alternative Possibility

Since this extraterrestrial theory has been discredited by so many people, is there an alternative possibility that would explain how these space beings visit our planet? The truth is that no extraterrestrial theory is credible, *except* where it can be interpreted in a spiritual context. Albrecht and Alexander state the case with clarity: "The extraterrestrial theory draws conclusions of a profoundly 'spiritual' nature while conveniently avoiding that controversial label. Entities that operate with total disregard for the inviolate laws of physics traveling at the speed of light or faster and having solved all their problems would have to be classified as spiritual semantic arguments notwithstanding."[53]

This change of climate has already begun. An increasing number of scientists and ufologists are coming over to this position, advocating what is known as the paraphysical hypothesis.

Even agnostic John Keel accepts this position. He states quite bluntly that the only hypothesis that can answer the questions posed by all the sightings is the paraphysical one. Other writers like R. A. F. Marshall Lord Dowding (who directed the *Battle of Britain* in 1940) came to the same conclusion. So did Dr. Clifford Wilson, when he wrote, "I (would) have no doubt whatever as to the fact of spiritual beings, evil forces, and phenomena that cannot be explained by purely physical, psychical or psychological concepts." In similar vein Vallee posed these questions to himself: "Why is it, I wondered that the "occupants" of UFOs behave so much like the denizens of fairy tales and the elves of ancient folklore? Why is the picture we can form of their world so much closer to the medieval concept of **Madonia**, the magical land above the clouds, than to a description of an extra-terrestrial planetary environment? And why are UFOs becoming a new religious form?"[54] He concludes: "The UFO phenomenon represents a manifestation of a reality that transcends our current understanding of physics...The UFOs are physical manifestations that cannot be understood apart from their psychic and symbolic reality. What we see in effect here is not an alien invasion. It is a control system which acts on humans and uses humans."[55]

The Condon Report

Before we proceed to examine this hypothesis in greater detail, it is interesting to note the various names that have been used by different authors to describe these visitors to our planet. Some refer to them as "humanoids," others as "ufonauts," "UFO sapiens," "paraphysical beings," and "spiritual beings." Scientists of late have shown a preference for this last designation and are referring more and more to these visitors as spiritual entities—*and indeed to the whole UFO activity as a spiritual or paraphysical phenomenon.* This was the term used in the Condon Report, the document published by an Air Force sponsored scientific group at the University of Colorado, under the direction of Dr. E. U. Condon (January, 1969). This report has been much criticized, and even ridiculed by certain Ufologists as another example of government whitewashing. But in fairness to the report, it should be noted that it did arrive at three significant conclusions:

(a) It all but demolished the idea that Earth was being visited by creatures from other planets and galaxies. Admitting that a few mysteries still remain unsolved, it concluded that there was no hard evidence that UFO space ships were from extra-terrestrial civilizations.

(b) The objects seen in the sky, and sometimes on the ground, were just as "material" as any other physical objects, although maybe only temporarily so.

(c) The sightings of UFOs, spaceships and aircraft and their occupants should be attributed to paraphysical phenomena. That is, UFOs belong to the same categories as *séances, spiritist movements, etc. In short, they are paraphysical.*

A Paraphysical Concept

As far as we know, this was the first time that the paraphysical concept had been advanced by a *group of scientists,* although individuals like Mead Layne had introduced the concept in the early 1950s. General Sanford of U.S. Intelligence had made a similar suggestion in 1952, comparing UFO sightings with spiritualism.[56] In 1959 the noted psychologist Carl Gustav Jung propounded the thesis that UFOs were not to be interpreted in physical terminology but rather in psychological and religious terminology. They should be understood not as interplanetary vehicles but "visionary" vehicles.[57]

Since 1959, there has been an accumulating mountain of evidence that would place UFO activity within a spiritual and religious framework. Many of the contactees when asked to describe what happened to them, referred to occult mysticism, psychic manifestations and telepathic communications. So pronounced have been the religious and cultic references, that a number of religious and semi-religious groups have crystallized around these alleged UFO sightings and communications. Some of these groups go by such names as the Aethesious Society, the Urautia Group, the Solar Cross Foundation, the School of Thoughts, plus hundreds of smaller and more secretive groups.

In spite of all that has been said about the skeptical attitude of the U.S. Government, it did publish an amazing statement in its preface to the official document "UFOs and Related Subjects: An Annotated Bibliography." It was composed by Miss Lynn E. Catoe in these measured terms: "A large part of the available UFO literature is closely linked with mysticism and the metaphysical. It deals with subjects like mental telepathy, automatic writing, and invisible entities, as well as phenomena like poltergeist manifestations and possession…many of the UFO reports now being published in the popular press recount alleged incidents that are strikingly similar to demonic possession and psychic phenomena which has long been known to theologians and parapsychologists."

This document was compiled for the United States Air Force and is now in the Library of Congress.

What exactly does one mean by the term "paraphysical"? And is it to be distinguished from the term "spiritual?"

The term "paraphysical" means that UFO vehicles and their occupants have the power to assume physical properties or solid, physical forms—and thus become visible to man; at the same time they have the power to become invisible again, and take on a spiritual or ethereal existence.

Dr. Clifford Wilson explains:

> Many of the reports of witnesses indicate that at least some UFOs were transparent, yet at the same time appearing to have mechanical or physical properties. There are strange stories of

> saucer occupants walking through the sides of their vehicles as they themselves were ghosts. Much of the evidence suggests the possibility that the vehicles themselves, and even the occupants of those vehicles, are not "permanent" solids in the sense that we would think of vehicles and beings of the earth. This is not to suggest that many of those vehicles are not solid at the time they are witnessed...For at the time they are sighted many of them are just as "solid" as houses and motor cars and other objects of earth all around us. They can leave impressions of their tripod stands on earth, and they can also be changed so that their molecular structure materializes or dematerializes in a way that we limited humans could not do.[58]

In his book UFOs...*Operation Trojan Horse,*[59] John Keel Writes: "The statistical data which I have extricated...indicate that flying saucers are not stable machines requiring fuel, maintenance, or logistical support. They are, in all probability, transmogrifications of energy and do not exist in the same way that this book exists. They **are not** permanent construction of matter." And again:

> All of these events seem to prove that a large part of the UFO phenomenon is hidden from us and has taken place beyond the limited range of our eyes. We can only see the entities and the object under certain circumstances **and perhaps only certain types of people can see them at all.**
>
> Thus by all standards of our sciences the UFOs do not really exist as solid objects. They may be a constant part of our environment but they are not an actual part of our reality."[60]

Physicist Jaques Lematre in the highly recognized *Flying Saucer Review*[61] writes: "We can consequently conclude that it is impossible to interpret the UFO phenomenon in terms of material spaceships as we conceive of the latter, i.e., in terms of manufactured, self-propelled machines retaining their material nature and their mechanical structure to travel from one solar system to another by traversing the distance separating these systems in the Einsteinian Continuum." The French physicist Vallee is quite emphatic in his conclusion: "They (UFOs) are constructed **as physical craft** (a fact which appeared to be undeniable) and as **psychic devices** whose exact properties remain to be defined."[62] And Dr. Hynek in an interview granted to *Fate Magazine* (June 1976) said:

> Perhaps an advanced civilization understands the interaction between mind and matter...Perhaps it is a naïve notion that you've got to build something physical, blast it off with sound and fury to cross vast distances and finally land here...There are other planes of existence—the astral plane, the etheric plane and so forth.
>
> I believe the world is in a psychic revolution that most of us are not aware of. And least aware are the establishment scientists...The new puzzle pieces are being given to us by the whole parapsychological scene—ESP, telepathy, the Uri Geller phenomena, psychic healing, and particularly psychic surgery.

There is an increasing number of scientists, however, who are attempting to explain and interpret the spirit world. As a result, a lot of new terminology has come into being; terms like "superspace," "space curving in on itself," "the edge of space," "something beyond the space-time-matter continuum." Some of these scientists have opted for the four-dimensional theory. Man, they say, lives in a world of three dimensions: length, width, depth. The fourth dimension is time. But man cannot control his passage through this dimension. He is time-bound. It is highly reasonable to assume that spirit-beings are not bound by matter or time.

One Day As a Thousand Years

Charles Camplejohn, scientist and former chief design engineer for the Apollo project, says that a strange phenomenon lies in the vortex of our galaxy:

> It's a rotating black hole. It's spinning around in one direction. If we were to get in our space ship and go into this black hole, funny things would happen. If we go with the rotation of the black hole, we will be going forward in time. If we go against the rotation, we will be going backward in time. It produces a paradox, because if you go with the rotation and then turn around and come back, you'll meet yourself coming.
>
> There's a peculiar thing associated with the rotating black hole and it's called a "singularity." If in our space ship we go past that singularity level, we have transferred time and space and have gone into another universe. This rotating hole is gobbling up stars at the rate of about thirty solar masses a year. That is a tremendous amount of energy. It's thirty times our sun each year, and the more stars that go into it each year the more mass is in there, and of course more energy...As the black hole sucks in these stars they pass the singularity level and are in another universe...They pass from universe to universe—forever. That is a wandering star.[63]

Scientist Camplejohn also has something to say on the question of travel outside of this black hole.

> Let me put you in a space ship and accelerate you to three-fourths the speed of light. Off you go into space while I stay here on earth. Two months later, by my calendar, you arrive

> back on earth. According to **your** calendar, you're been gone only **one** month. I'm two months older and yet you've aged just one month.
>
> Let's accelerate you a little more—99.9 percent the speed of light—and send you off again. This time when you get back to the earth you've still aged just thirty days, while I've aged ten years. The reason for this is that moving clocks run slower than stationary clocks when they're taken in reference to their inertial time frame.
>
> Now let's crank your speed up to 99.999999 percent the speed of light. This time we send you off into space headed for the center of our galaxy, some 27,000 light years away. You would reach the center of our galaxy in twenty-seven days. The earth in the meantime would have aged 27,000 years. There you have it. **One day as a thousand years!** And Peter wasn't a physicist or an astronaut. (2 Peter 1:21)

Then he adds this intriguing comment: "We said...We had gone up to 99.999999 percent the speed of light. There is a good reason why I never said we reached the speed of light. It has to do with the square root of minus. If it were possible for man to reach the speed of light, matter ceases to exist. What phenomenon exists beyond that point we don't really know."

Travelling in Four Dimensions

This opens up a whole lot of possibilities. If these spirit-beings could travel in all four dimensions, it would explain a number of things about angelic appearances and disappearances. But supposing there are more than four dimensions. Supposing there are five, or six, or even seven, and they could travel in all of them! Then nothing would be impossible.

Perhaps a more homely illustration is the one suggested by Alan Landsburg in his book, "In Search of Ancient Mysteries." He discusses the possibility of short-cutting through space-time in a multi-dimensional universe:

> In New York are two apartment buildings, back to back. The entrance to one is on Fifth Avenue, the entrance to the other on Sixth. Mr. White and his wife live on the seventh floor of one of those buildings. A wall of their living room is the back wall of the building.
>
> Their friends the Blacks live on the seventh floor of the other building, and a wall of **their** living room is the back wall of their building. So these two couples live within two feet of each other, since the back building walls actually touch. But, of course, they don't see or hear each other.
>
> When the Blacks want to visit the Whites, they walk from their living room to the front door. Then they walk down a long hall to the elevator. They ride seven floors down. Then, in the street, they must walk around to the next block—and the city blocks are long. In bad weather they must sometimes actually take a cab. They walk into the other building, they go through the lobby, ride up seven floors, walk down a hall, ring a bell, and finally enter their friends' living room—only two feet from their own.
>
> The way the Blacks travel is like our own civilization's space travel—the actual physical crossing of enormous three-dimensional spaces. But if they could only step through those two feet of wall without harming themselves or the wall—well, maybe that's how the old ones come here from their mysterious planet.[64]

Since we have no way of proving or disproving these hypothesis, we are left with speculation. Science seems to have exhausted its resources. This being so, one is justified in asking whether biblical revelation has any additional data to supply. Does it have information to give that science so far has been unable to ascertain. This brings us to our leading question: Does the Bible reveal any knowledge of spirit-beings? Does it speak of extraterrestrial visitors to planet Earth?

5

"Sons of God," "Daughters of Men"

The Craving of demons for a body,
evident in the Gospels, offers at least some
parallel to this hunger for sexual experience.

—Derek Kidner

In 1947 an Arab boy tending his sheep accidentally discovered an ancient cave near the Dead Sea. In it were found a priceless collection of ancient scrolls which soon became known as the Dead Sea Scrolls or the Qumran Texts. Among these writing was one known as the Genesis Apocryphon. At first it was thought to be the long lost book of Lamech. Although the scroll consisted of a speech by Lamech and a story about some of the patriarchs from Enoch to Abraham; it was not that book.

According to the Bible, Lamech was the son of Methuselah and the father of Noah. He was the ninth of the ten patriarchs of the antediluvian world.

It is significant, however, that the Genesis Apocryphon mentions the Nephilim, and makes reference to the "sons of God" and the "daughters of men" introduced in Genesis 6. The Apocryphon also elaborates considerably on the succinct statements found in the Bible, and provides valuable insights into the way these ancient stories were interpreted by the ancient Jews.

The copy of the Genesis Apocryphon discovered at Qumran dates back to the second century BC, but it was obviously based on much older sources. When discovered in 1947, it had been much mutilated from the ravages of time and humidity. The sheets had become so badly stuck together that years passed before the text was deciphered and made known. When scholars finally made public its content, the document confirmed that celestial beings from the skies had landed on planet Earth. More than that, it told how these beings had mated with Earth-women and had begat giants.

Is this story myth or history, fable or fact? Specialized research has revealed that many ancient legends have a basis in fact. But to answer the question, let us consult the most authoritative document known to man—the Bible.

In Genesis 6:1–4 the "sons of God" are captivated by the beauty of the "daughters of men." They subsequently marry them and produce an offspring of giants known as the Nephilim. Genesis goes on to say that these Nephilim were "mighty men" and "men of renown."

"Sons of God"? "Daughters of men"? What sort of beings were these? Were they human or did they belong to an alien species from outer space?

Identifying the Sons of God

There is no problem in identifying the "daughters of men" for this is a familiar method of designating women in the Bible. The problem lies with the "sons of God." Three major interpretations have been offered to shed light on this cryptic designation.

First, a group within orthodox Judaism theorized that "sons of God" meant "nobles" or "magnates." Hardly anyone today accepts this view.

Second, some interpret the "sons of God" as fallen angels. These were enticed by the women of Earth and began lusting after them. Many reputable Bible commentators have rejected this theory on psychophysiological grounds. How can one believe, they ask, that angels from heaven could engage in sexual relations with women from Earth? Philastrius labeled such an interpretation a down-right heresy.

Third, many famed scholars contend that the "sons of God" are the male descendants of Seth, and that the "daughters of men" are the female descendants of Cain. According to this view, what actually happened in Genesis 6 was an early example of believers marrying unbelievers. The good sons of Seth married the bad daughters of Cain, and the result of these mixed marriages was a mongrel offspring. These later became known for their decadence and corruption; indeed, it reached such a degree that God was forced to intervene and destroy the human race. This comment of Matthew Henry could be taken as representative of those holding the view: "The sons of Seth, (that is the professors of religion) married the daughters of men, that is, those that were profane, and strangers to God and godliness. The posterity of Seth did not keep by themselves, as they ought to have done. They intermingled themselves with the excommunicated race of Cain."[65]

However, in spite of the excellent pedigree of the proponents of this theory, their argument is not convincing. Their interpretation is pure eisegesis—they are guilty of reading into the text what is obviously not there.

False Exegesis

Their interpretation fails on other grounds as well. At no time, before the Flood or after, has God destroyed or threatened to destroy the human race for the sin of "mixed marriages." It is impossible to reconcile this extreme punishment with the mere verbal strictures found elsewhere in the Bible for the same practice. If God is going to be consistent, He should have destroyed the human race many times over!

The contrast made in Genesis 6:2 is not between the descendants of Seth and the descendants of Cain but rather between the "*sons* of God" and the "*daughters* of men." If by "*sons* of God" is meant "sons of Seth," then only the sons of Seth engaged in mixed marriages, and not the daughters. And only the *daughters* of Cain were involved, and not the sons. And another strange assumption is implied: that only the *sons* of Seth were godly, and only the *daughters* of Cain were evil.

The strangeness is compounded when one seeks for evidence that the sons of Seth were godly. We know from Genesis that when the time came for God to destroy the human race, He found only one godly family left among them—that of Noah. Where were all the other supposedly godly sons of Seth? Even Seth's own son could hardly be called righteous. His name was Enos, meaning "mortal" or "frail." And he certainly lived up to it! Genesis 4:26 reads, "And to Seth, to him also there was born a son; and he called his name Enos: *then began men to call upon the name of the Lord*" (emphasis added). That statement seems harmless enough, but what does it mean when it says that it was only now that men began to call upon the name of the Lord? Upon whom did Adam call? And Abel? And Seth himself?

Some scholars give us a more literal and exact translation to this verse: "Then men began to call themselves by the name of Jehovah." Other scholars translate the statement in this manner: "Then men began to call upon their gods (idols) by the name of Jehovah." If either of these be the correct translation then the evidence for the so-called godly line of Seth is non-existent. The truth of the matter is that Enos and his line with few noted exceptions were as ungodly as the other line. The divine record could not be clearer: "*all flesh* had corrupted his way upon the earth" (Genesis 6:12; emphasis added).

In the Old Testament, the designation "sons of God" (bene Elohim) is never used of humans, but always of supernatural beings that are higher than man but lower than God. To fit such a category only one species is known—angels. And the term "sons of God" applies to both good and bad angels. These are the beings of whom

Augustine wrote: "Like the gods they have corporeal immortality, and passions like human beings."[66]

The designation "sons of God" is used four other times in the Old Testament, each time referring to angels. One example is Daniel 3:25, where King Nebuchadnezzar looks into the fiery furnace and sees four men, "and the form of the fourth is like the son of God." The translation is different and clearer in our modern versions, "like a son of the gods." Since Jesus had not yet become the "only begotten son" of God, this "son" would have to be angelic.

Another example is Job 38:7, which says the sons of God shouted for Joy when God laid the foundations of the Earth. Angels are the only entities that fit this designation since man had not been created at that time!

In Job 1:6 and Job 2:1 the "sons of God" came to present themselves before the Lord in Heaven. Among the sons of God is Satan—a further implication that the "sons of God" must have been angels.

Since the designation "sons of God" is consistently used in the Old Testament for angels, it is logical to conclude that the term in Genesis 6:2 also refers to angels.

Sons of God: Three Categories

In the New Testament, born-again believers in Christ are called the children of God or the sons of God (Luke 3:38, John 1:12, Romans 8:14, 1 John 3:1). Dr. Bullinger in the Companion Bible states: "It is only by the divine specific creation that any created being can be called a 'son of God.'" This explains why every born-again believer is a son of God. It explains also why Adam was a son of God. Adam was specifically created by God, "in the likeness of God made He him" (Genesis 5:1). Adam's descendants, however, were different; they were not made in God's likeness but in Adam's. Adam "begat a son in his own likeness, after his image" (Genesis 5:3). Adam was a "son of God" but Adam's descendants were "sons of men."

Lewis Sperry Chafer expresses this in an interesting way when he states: "In the Old Testament terminology angels are called sons of God while men are called servants of God. In the New Testament this is reversed. Angels are the servants and Christians are the Sons of God."[67]

It is thus clear that the term "sons of God" in the Bible is limited to three categories of beings: angels, Adam and believers. All three are special and specific creations of God. As for the use of the term in Genesis 6, since it cannot possibly refer to Adam nor believers in Christ, we conclude that it has to refer to the angels whom God had created.

Light from the New Testament

Two New Testament passages shed further light on Genesis 6. They are Jude 6–7 and 2 Peter 2:4. These verses indicate that at some point in time a number of angels fell from their pristine state and proceeded to commit a sexual sin that was both unusual and repugnant. Jude 6–7 states: "And the angels which kept not their first estate, but left their own habitation, he hath reserved in everlasting chains under darkness unto the judgement of the great day. Even as Sodom and Gomorrah and the cities about them in like manner, giving themselves over to fornication, and going after strange flesh..."

These angels not only failed to keep their original dominion and authority, but they "left their own habitation." Habitation is a significant word: it means "dwelling place" or "heaven." And the addition of the Greek word *idi* (their own) means that they left their own private, personal, unique possession.[68] Heaven was the private, personal residence of the angels. It was not made for man but for the angels. This is why the ultimate destination of the saints will not be Heaven but the new and perfect Earth which God will create (Revelation 21:1–3). Heaven is reserved for the angels, but as for the beings referred to in Jude 6–7, they abandoned it.

Not only did these angels leave Heaven, they left it once-for-all. The Greek verb *apoleipo* is in the aorist tense, thus indicating a once-for-all act. By taking the action they did, these angels made a final and irretrievable decision. They crossed the Rubicon. Their action, says Kenneth Wuest, "was apostasy with a vengeance."[69]

As to the specific sin of these angels, we are given the facts in Jude 7. As in the case of Sodom and Gomorrah it was the sin of "fornication" and "going after strange flesh." "Strange" flesh means flesh of a different kind (Greek *heteros*). To commit this particularly repugnant sin, the angels had to abandon their own domain and invade a realm that was divinely forbidden to them. Says Wuest: "These angels transgressed the limits of their own natures to invade a realm of created beings of a different nature."[70] Alford confirms: "It was a departure from the appointed course of nature and seeking after that which is unnatural, to other flesh than that appointed by God for the fulfillment of natural desire." The mingling of these two orders of being, was contrary to what God had intended, and summarily led to God's greatest act of judgement ever enacted upon the human race.

Tempting the Angels

Another New Testament verse that may have bearing on Genesis 6. In 1 Corinthians 11:10, Paul instructs that a woman should cover her head as a sign of subjection to her husband, and also *"because of the angels"* (emphasis added). This observation has intrigued commentators across the years. Why this sudden reference to angels? Could it be a reference to what happened in Genesis 6 where angels succumbed to the inducements and physical charm of the women of Earth? Obviously, Paul believed that an uncovered woman was a temptation even to angels. William Barclay mentions an old rabbinic tradition which alleges that it was the beauty of the woman's long hair that attracted and tempted the angels in Genesis 6.

Strange Parentage

The offspring of this union between the "sons of God" and the "daughters of men" were so extraordinary that it indicates an unusual parentage. In no way could the progenitors of such beings be ordinary humans. Their mothers possibly could be human, or their fathers, but certainly not both. Either the father or the mother had to be superhuman. Only in such a way can one account for extraordinary character and prowess of the off-spring.

God's law of reproduction, according to the biblical account of creation is "everything after his kind." God's law makes it impossible for giants to be produced by normal parentage. To produce such monstrosities as the Nephilim presupposes supernatural parentage.

Giants?

"Nephilim" is a Hebrew word translated in the Authorized King James Version as "giants." "There were giants in the earth in those days" (Genesis 6:4). It is true that they were giants in more senses than one. However, the word Nephilim does not mean "giants." It comes from the root *naphal,* meaning "fallen ones," and most modern versions of the Bible have left the word "Nephilim" untranslated.

When the Greek Septuagint was made, "Nephilim" was translated as *gegenes.* This word suggests "giants" but actually it has little reference to size or strength. *Gegenes* means "earthborn." The same term was used to describe the mythical "Titans"—being partly of celestial and partly of terrestrial origin.[71]

The Hebrew and the Greek words do not exclude the presence of great physical strength. Indeed, a combined supernatural and natural parentage would imply such a characteristic. Angels, according to scripture, are known for their power. They are often referred to as "sons of the Mighty" (Psalm 103:20). Therefore, if the ones who sired them were strong and mighty, it could be assumed that their offspring were likewise.

No evidence exists in Scripture that the offspring of mixed marriages (believers and unbelievers) were giants, excelling in great strength and might. No evidence can be found anywhere in history for that matter. Such an interpretation poses impossible assumptions.

When the word "Nephilim" is used in Numbers 13:33, the question of size and strength is explicit. Here we are left in no doubt as to their superhuman prowess. When Joshua's spies reported back from Canaan, they called certain of the inhabitants of Canaan "giants." "And there we saw the Nephilim, the sons of Anak, which come of the Nephilim, and we were in our own sight as grasshoppers, and so we were in their sight."

Some commentators have speculated that the Nephilim of Numbers 13 belonged to a second eruption of fallen angels, since the earlier Nephilim had been destroyed in the Flood. And they see an allusion to this in Genesis 6:4, where it states that "there were Nephilim in the earth in those days; *and also after that,* when the sons of God came in unto the daughters of men" (emphasis added). Could it be that the "after that" was a reference to the Nephilim found in Canaan during the Israelite entry into the land? If so, it could explain why the Lord commanded the total extermination of the Canaanites, as He had earlier ordered the near annihilation of the human race.

Nephilim—No Resurrection

The book of Isaiah says that the Nephilim and their descendants will not participate in a resurrection as is the portion of ordinary mortals. Isaiah 26:14 reads: "They are dead, they shall not live; they are deceased, they shall not rise." The original Hebrew word translated "deceased" here is the word "Rephaim." It would have saved a lot of misinterpretation of the translators had left the word as it was in the original. The verse actually reads: "Dead, they shall not live; Sephaim, they shall not rise." The Rephaim are generally understood to be one of the branches of the Nephilim, and God's Word makes it clear that they are to partake in no resurrection. But with humans it is different: all humans will be resurrected either to life or to damnation (John 5:28-29).

We have already seen that the Greek Version of the Old Testament (The Septuagint) translated "Nephilim" as *gegenes*; we shall now inquire how it translates "sons of God." In some of the manuscripts it is left as "sons of God," but in the others—including the Alexandrian text—it is rendered by the word *angelos*. This text was in existence in the time of Christ, but there is no indication that He ever corrected or queried it. Can we not assume from His silence that He agreed with the translation!

Rape of the Text

Having studied all the arguments in favor of "sons of Seth," one concludes that the only argument that is valid among them is that of rationality. "Sons of Seth" is an interpretation that is more palatable to human reason. Reason can never subscribe to the incredible notion that fallen angels could have sex relations with women of Earth. Angels have no physical bodies! They do not marry! They belong to an entirely different species of being! The mind revolts against such absurdity. So, what does one do? Settle, of course, for an easy, rational interpretation—sons of Seth and daughters of Cain. But what if the meaning of Scripture is clearly otherwise? There is the rub! Scripture is clearly otherwise! To impose a human interpretation at the expense of the obvious meaning of the divine Word, is a rape of the biblical text. Furthermore, when one deals with the world of the supernatural, rationality is never an argument.

Jewish and Patristic Fathers

The Jewish Fathers, when interpreting this expression from Genesis 6:2, invariably interpreted it as "angels." No less an authority than W. F. Allbright tells us that: The Israelites who heard this section (Genesis

6:2) recited unquestionably thought of intercourse between angels and women.[72] Philo of Alexandra, a deeply religious man, wrote a brief but beautiful treatise on this subject, called "Concerning the Giants." Basing his exposition on the Greek version of the Bible, he renders it as "Angels of God." Says Bamberger, "Had he found the phrase 'sons of God' in his text, he most certainly would have been inspired to comment on it."[73]

Philo certainly took the Genesis as historical, explaining that just as the word "soul" applies to both good and evil beings, so does the word "angel." The bad angels, who followed Lucifer, at a later point in time failed to resist the lure of physical desire, and succumbed to it. He goes on to say that the story of the giants is not a myth, but it is there to teach is that some men are earth-born, while others are heaven-born, and the highest are God-born.[74]

The Early Church Fathers believed the same way. Men like Justin Martyr, Irenaeus, Athenagoras, Tertullian, Lactantius, Eusebius, Ambrose...All adopted this interpretation. In the words of the Ante-Nicene Fathers, the angels fell "into impure love of virgins, and were subjugated by the flesh..."Of those lovers of virgins therefore, were begotten those who are called giants."[75] And again, "...The angels transgressed, and were captivated by the love of women and begat children."[76]

Nowhere before the fifth century AD do we find any interpretation for the "sons of God" other than that of angels. We cannot deny the Jewish Fathers knowledge of their own terminology! They invariably translated "sons of God" as "angels." The testimony of Josephus, that colorful cosmopolitan and historian, is also of paramount importance. In his monumental volume, *Antiquities of the Jews,* he reveals his acquaintance with the tradition of the fallen angels consorting with women of Earth. He not only knew of the tradition but also tells us how the children of such union possessed super human strength and were known for their extreme wickedness. "For the tradition is that these men did what resembled the acts of those men the Grecians called giants." Josephus goes on to add that Noah remonstrated with these offspring of the angels for their villainy.[77]

Perhaps the most conclusive argument for interpreting the expression as "angels" is the simplest one of all. If the writer of Genesis wanted to refer to the "sons of Seth" he would have just said so. If God had intended that meaning, then the verse would undoubtedly read, "those sons of Seth saw the daughters of Cain that they were fair…" But the Bible meant something far more sinister—the sexual union between angels from Hell and evil women from Earth. Because of the gravity of such a union, and its dire consequences for the human race, God moved to destroy the race before it could destroy itself—except for one family which had not been contaminated.

The Ultimate Sin

God made man in His own image, the highest of all His earthly creations. While God said that everything He made was good, He considered man *very* good. Man had been made for fellowship with God himself, but he soon turned his back upon His maker and worshipped the creature more than the Creator. Before many generations, the human race was being polluted by this abominable union with demons. It seemed that Hell and Earth were in league together against the God of Heaven. God's righteous anger was such that He regretted having made man. "And God saw that the wickedness of man was great in the earth, and that every imagination of the thoughts of his heart was only evil continually. And it repented the Lord that he made a man…" (Genesis 6:5–6). It was specifically because of this ultimate sin that God brought about a deluge of such magnitude that man and beast were drowned from the face of the Earth. In the words of old Joseph Hall: "The world was so grown foul with sin, that God saw it was time to wash it with a flood: and so close did wickedness cleave to the authors of it, that when they were washed to nothing, yet it would not wash off; yea, so deep did it stick in the very grain of the earth, that God saw it meet to let it soak long under the waters."[78]

Why Was Noah Immune?

Why Noah and his immediate family were the only ones immune from this great judgement is significant. Genesis 6:9 says, "Noah was a just man." He stood out as an example of righteousness and godliness in a perverse age. Like Enoch before him, Noah also "walked with God." But there was another reason why Noah was spared, one that seems to have escaped most commentators. Genesis 6:9 says that Noah was "perfect in his generation." Does this mean moral and spiritual perfection? Hardly. Genesis 9:20–23 disproves any such perfection. What, then, does the Bible mean by calling him "perfect?" The Hebrew word is *tamiym* and comes from the root word *Taman*. This means "without blemish" as in Exodus 12:5, 29:1, Leviticus1:3. Just as the sacrificial lamb had to be without any physical blemish, so Noah's perfection. In its primary meaning, it refers not to any moral or spiritual quality but to physical purity. Noah was uncontaminated by the alien invaders. He alone had preserved their pedigree and kept it pure, in spite of prevailing corruption brought about by the fallen angels.[79] And again: "Noah's bloodline had remained free of genetic contamination."[80] This implies, of course, that all other families on Earth had been contaminated by the Nephilim. It also proves that the assault of Satan on the human race had been far more extensive than realized. It is no wonder that God pronounced such a universal fiat of judgement.

As for the fallen angels who participated in the abomination, God put them in custody "in everlasting chains under darkness unto the judgement of the great day" (Jude 6). This is sometimes interpreted as Tartarus or the "nether realms" (2 Peter 2:4). This would also explain why some fallen angels are in custody and why others are free to roam the heavens and torment mankind.

Such a drastic punishment, both for men and angels, presupposed a drastic sin, something infinitely more evil and more sinister than mixed marriages. It was nothing less than the demonic realm attempting to pervert the human world. By genetic control and the production of hybrids, Satan was out to rob God of the people He had made for himself.

If Satan had succeeded in corrupting the human race, he would have hindered the coming of the perfect Son of God, the promised "seed of the woman," who would defeat Satan and restore man's dominion (Genesis 3:15). If Satan had by any means prevented that birth, he would obviously have averted his own doom. Satan did succeed to a large extent. It was for this reason that God drowned mankind in the Deluge.

Are Angels Sexless?

Interpreting the "sons of God" as fallen angels, the question immediately arises—*do angels marry?* In Matthew 22:30, Jesus said angels neither marry nor are given in marriage. This seems a clear and emphatic negative. However, it does not preclude the possibility of such a thing happening—obviously contrary to the will of God. And it does not preclude fallen angels, who had rebelled against God already, from cohabiting with women of Earth, as the Scriptures state.

Some interpret the words of Jesus as meaning that angels do not marry *among themselves.* Is it because they are all male? Or is it because celestial beings are deathless and thus need no offspring. Only terrestrial beings need to find immortality in their children.[81] But if they do not need to marry and procreate, is it still possible that they could engage in sexual acts? If not among themselves then with human spouses? Jude seems quite explicit on the matter: the angels left their own habitation and gave themselves over to fornication, going after strange flesh. In other words, they were capable of performing human functions—eating, drinking, walking, talking, even sexual activity and fathering children.

The fact that angels do not marry does not in itself prove that they are sexless. Throughout the Bible, angels are referred to only as *men.* Finis Drake writes: "It is logical to say…that the female was created specifically for the human race in order that it could be kept in existence; and that all angels were created males, in as much as their kind is kept in existence without the reproduction process. Angels were created innumerable to start with (Hebrews 12:22) whereas, the human multitudes began with one pair."[82]

Even in the next world, when the saints will dwell in their resurrection body and live forever, it does not imply that they will be sexless. The Bible teaches that everyone will have *his own body* in the resurrection (1 Corinthians 15:35–38). No suggestion is made that they will be unsexed. Furthermore, Christ remained a man after His resurrection.

Demons at Large

One other question has been raised. If the fallen angels who lusted after women of Earth in Genesis 6 have been interred in Tartarus with "everlasting chains" how does one explain the demons who have been operating since then? They seemed to have been quite active during the ministry of Jesus, and are busy again in our day. Following this reasoning, some share the conclusion of Kent Philpott: "However one might wish to interpret Genesis 6:1–4 to link this passage with the verses in 2 Peter and Jude seems to post far more problems than it would solve. But 2 Peter 2–4 and Jude clearly assert that the rebellious angels are being kept prisoner in the "nether gloom." If they are prisoners, they could not very well function as the demons are described as functioning in the New Testament."[83] But Philpott failed to see that there are two categories of fallen angels: Those cast out of Heaven with Lucifer, and who are still free to torment mankind; and those who fell the second time by committing carnal acts with the daughters of men. The spirits in this second category are those chained in the nether regions.

It seems clear to me that the "sons of God" are none other than fallen angels, and because of their further sin of lusting after the "daughters of men" many were imprisoned by God. Both the near annihilation of the human race and the incarceration of the fallen angels in Tartarus indicate the magnitude of the sin they committed. By such drastic judgement, God saved the human race from a calamity worse than the physical death originally imposed upon them.

6

The Evidence Mounts

More and more we are finding that mythology
in general though greatly contorted very often
has some historic base. And the interesting thing
is that one myth which occurs over and over
again in many parts of the world is that somewhere
a long time ago supernatural beings had sexual intercourse
with natural women and produced a special breed of people.

—Francis A. Schaeffer

Evidence for the existence of the Nephilim goes beyond the biblical record. The story of Lamech, found in the Genesis Apocryphon, relates how Lamech had been away from home on a long journey. When he finally came back, he discovered to his chagrin that his wife, Bat-Enosh,[84] had given birth to a baby boy in his absence. He was sure that the child had not been sired by him, and what is more, the child bore no resemblance to him or to anyone else in the family. Adding to the mystery was the fact that the boy was extremely beautiful, and when he opened his eyes he lighted up the whole house. "I have begotten a strange son," said Lamech, "...His nature is different, and he is not like us, and his eyes are as the rays of the sun, and his countenance is glorious. And it seems to be that he is not sprung from me but from the Angels..."[85]

Lamech did what most husbands would have done: he reproached his wife for infidelity. Bat-Enosh, however swore by all that was sacred that Lamech himself must have fathered the child. She had not known any other man, not a stranger—and note this—not a *Watcher or Heavenly Being.*

How enlightening to our study! Who were these Watchers or Heavenly beings? According to the Book of Daniel, they were fallen angels (Daniel 4:13, 17, 23).

Here is a statement made by Bat-Enosh as it reads in the Lamech scroll: "My lord and kinsman, remember my delicate feelings. How (ever), the occasion is indeed alarming, and my soul (is writhing) in it sheath. I will tell you everything truly."

Then she saw how perturbed her husband was, and decided to repress her passion and indignation a little: My lord and kinsman, (I will ignore) my delicate feelings and swear to you by the Holy (and) Great One, the Sovereign of heaven (and earth) that this seed came from you, and this fruit was planted by you and not by some stranger or by any of the **Watchers or heavenly beings.** (Have done with) this troubled and marred expression and this gloomy mood. I am telling you the truth.[86]

Lamech by this time must have begun to realize that the child born *could* have been conceived by one of these *Watchers or Heavenly Beings.* If so, his child belonged to the Nephilim.

Mystery of Noah's Birth

Not completely sure as to what to believe, Lamech sought the advice of his father on the matter. Methusnnnaleh listened attentively as Lamech revealed this strange story, then promised his son that he would seek the advice of his father, the wise and godly Enoch. Since the family's reputation was at stake, something had to be done.

Enoch, whose name meant "the intelligent" or "the learned," sensed the meaning of what had happened. He sent Methuselah home with the disturbing news that the Earth would soon be visited by a terrible catastrophe and judgement. It was clear that corruption had taken place, and that the human race had become tainted. God would soon be moving in judgement, and human flesh would perish. And as for this little boy, whose birth remained a mystery, he should be raised by Lamech, and should be called Noah. What is more, little Noah had been specially chosen by God to survive this coming judgement, and would be the progenitor of the new inhabitants of planet Earth.

Despite the mystery surrounding his birth, Noah could not be one of the Nephilim, as we shall see in a later chapter. There is no doubt, however, that the unusual circumstances of his birth convinced Enoch, Methuselah, Lamech and Bat-Enosh that strange things were happening on the Earth.

The Genesis Apocryphon is not the only extra-biblical documentation. Numerous other documents indicate the existence of the Nephilim. I do not place these documents on par with Scripture, nevertheless they seem to corroborate Scripture. My sole purpose for introducing them is simply to state that they exist and that they share common features with the Scriptures. My task is similar to that of Montaigne: "Gentlemen, all I have done is make a bouquet of flowers from flowers already picked, adding nothing but the string to tie them together."[87]

AN ANCIENT BEST SELLER

One such document is the Book of Enoch. This is a composite volume, and, although not known in Europe until the eighteenth century, it was a veritable best-seller in the days of Christ. In the centuries immediately preceding and following the coming of Christ, this book was widely read and discussed, and its impact was tremendous. Without question it is the most notable apocalyptical work outside the canonical Scriptures. R. H. C. Charles, an outstanding authority in this field, tells us that "the influence of 1 Enoch on the New

Testament has been greater than that of all the other apocryphal and pseudepigraphical books put together."

For one thing, the Book of Enoch gave to the world the concept of a pre-existent Messiah, and by so doing prepared the way for Christian doctrine. It was from this same book that the Manual of Discipline (found at Qumran) received its solar calendar. What is more, this book became so influential that it became an exemplar and a catalyst for the burgeoning apocalyptic literature of the time. Indeed, it can be claimed, that the Book of Enoch was one of the most important apocalyptic books ever written.

Tertullian and some of the other Church Fathers considered it of such import that they included it as part of the sacred canon of Scripture. Jude 14 is a direct quote from Enoch 1:9, 5:4, and 27:2. There is also in the Book of Enoch a doctrine that one finds nowhere else. Other authors tell us that each nation has its sar or parton, and that of Israel was Michael, or sometimes God Himself. However, in the Book of Enoch we are told that "*God* was Israel's shepherd till the last years of the Kingdom of Judah; then in disgust He turned them over, not to their own guardian, but to the sarim of the Gentiles."[88]

An amazing book! Such a doctrine prefigures and predicts the "times of the Gentiles" mentioned in the Bible. Yet, of all mysteries, this book remained neglected for more than 1,800 years, and continues to remain neglected, and even shunned, in the twentieth century.

ENOCH

Who was this Enoch, whose name is attached to the book? Enoch is mentioned in the Scriptures, but we are told little about him. Just four verses in all—two in the Old Testament and two in the New. That little, however, is sufficient to distinguish him as one of the outstanding men of all times. He is listed as the seventh of the ten patriarchs between Adam and Noah. He was the father of

Methuselah, the man holding the world's record for longevity. Interestingly, Enoch became a believer and started "walking with God" after his son was born. Soon he became known for his exceptional piety and godliness. The most remarkable fact of all is that he did not die. "Enoch walked with God; and he was not, for God took him." He is one of the only two men mentioned in the Bible who were translated to Heaven without tasting death.

In the Book of Enoch other significant facts are given about this patriarch. It claims that Enoch will return to Earth at the end of time. And that he will be one of the two martyrs (or witnesses) slain on the streets of Jerusalem.

The book of Yasher adds still more detail. It tells how Enoch would periodically withdraw himself from earthly company, and visit the ten heavens. He would then return to Earth with a divine luster on his face, as Moses emerged from the presence of God on Mt. Sinai, and knew not that his face shone. Such was the regard in which Enoch was held by some of the Jewish mystics that they referred to him as "the lesser Yahweh"—a title bordering on blasphemous.

In the Book of Jude (verses 14–15), we are told of a prophecy Enoch made concerning the coming of the Lord. This may well be the oldest literary statement in existence. In it, Enoch predicts: "Behold, the Lord cometh with ten thousand of his saints, to execute judgement upon all, and to convince all that are ungodly among them of all their ungodly deeds which they have committed, and of all their hard speeches which ungodly sinners have spoken against him."

It is obvious that Jude had access to the Book of Enoch, and he did not hesitate to quote from it.

Most scholars claim that the Book of Enoch could not have been written by the Enoch of the Bible. Because of this, the book is labeled a pseudepigrapha. What happened, they suggest, is that a much later author complied the work—probably around the second century

BC—and used a number of different sources. The author then added the name of the biblical Enoch. In this way, he was assured of the book's acceptance! Whether or not this was so, is not our present concern. But even if it were true, it is obvious that the compiler of the book must have used ancient sources that were at his disposal. He was not so much the author as the compiler or redactor of the book. What could be more honest, more modest, than for a compiler to attach to his book not his own name but that of his oldest source. And who older than Enoch?

An Ancient Astronaut

Enoch obviously had access to knowledge and information completely beyond the reach of mortal man at that stage in man's development. Some of this knowledge couches in terminology that is allegorical and symbolical, and its meaning frequently escapes us. However, beneath the abstruse verbal descriptions there is some amazing data. The fact that it is conveyed in imaginative and allegorical terminology should not be allowed to detract from its basic accuracy.

The important question in considering the Book of Enoch is whether it contains any reference to the strange and bizarre event recorded in Genesis 6? Does it have anything to say about the "sons of God" and the "daughters of men" coming together in sexual union? Yes, it does, many times. It contributes a number of details not found in the Genesis record. But none of these facts contradict in any way what is told us in Genesis. Instead, they elaborate on the facts already given. What, then, is this additional information?

One item of interest is Enoch's account of a visit to the "fifth heaven," where he saw giants with their "faces withered, and the silence of their mouth perpetual." Enoch calls them the "Grogori" or "fallen angels," who broke their vows, married the daughters of men, and "befouled the earth with their deeds." He also mentions that "giants were born and marvelous big men and enmity."

The Watchers

Significantly, the Book of Enoch (like the Genesis Apocryphon) refers to these "sons of God" as *Watchers.* A term, as already noted, found in the Book of Daniel.[89] Why this particular word should be used, we are not told; possibly because part of their function was to keep vigil. R. H. C. Charles, who translates the word "watchers" as "the angels, the children of heaven," obviously believed they referred to one and the same beings.

The Book of Enoch alleges that two hundred of these *Watchers* descended to Earth in the days of Jared (Genesis 5:18), and some of them are given names. The worst one of all was called Azazel. The name occurs in other Jewish documents, like the Apocalypse of Abraham. Azazel is accused of having "scattered over the earth the secrets of heaven and hath rebelled against the Mighty One." His name is also found in ancient Jewish ritual concerning the Day of Atonement. On that day, the iniquities of the people of Israel were laid on the scapegoat, and then the scapegoat was driven away "to Azazel, to the wilderness" (Leviticus 16). Azazel was a demon who inhabited a region in the Judean wilderness.

There is an interesting aside on Azazel in a brilliant essay by Dr. Jacob Z. Lauterbach explaining certain "stray references to the activity of Satan on Yom Kippur." The ritual of that day, said Lauterbach, sought to negate Satan's efforts in three ways: The Azazel sacrifice to appease him; the smoke of the incense to drive him away; and the whiter robes of the High Priest, so different from his usual vestments, a disguise to mislead him.[90]

In the Book of Jubilees a different reason is given as to why the Watchers came to Earth. It was in order "to instruct the children of men and to bring about justice and equity on the Earth." However, the story ends in the same way: instead of instigating justice and equity, they lusted after the women of Earth, and merited the full judgement of God.

Secrets of the Cosmetic Trade

According to the Book of Enoch, these Watchers instructed the people of Earth in many studies, including the use of charms and enchantments, the arts of magic, and the secrets of the cosmetic trade. And the Book of Jubilees adds to the list, "a medical herbal" which Noah wrote down at the dictation of the angels.

It is interesting how the early Church Fathers used some of these details for preaching purposes and moralistic instruction. Foremost among these was Tertullian, and certainly one of the most original. He asks with astringent irony, why should the angels have to instruct the women of Earth in cosmetic arts? Surely, if they had succeeded in charming angels without any cosmetic aids, it were an easy matter to charm men!

Astrology, Weapons, Et Al

Another study in which the Watchers instructed earthlings was astrology, with all its concomitant evils. And last but not least, they instructed men how to fashion weapons of destruction, particularly swords, knives and shields. Comyns Beumont deduces that they also made ammunition and even explosives.[91]

The Book of Enoch confirms the Book of Genesis to the letter when it states: There arose much godlessness, and they committed fornication, and they were led astray, and became corrupt in all their ways.[92]

Things got so bad on planet Earth, according to the Book of Enoch, that the archangels of heaven—Michael, Uriel, and Gabriel—reported the matter to the Most High God: The whole earth has been filled with blood and unrighteousness. And now, behold, the souls of those who have died are crying and making their suit to the gates of heaven, and their lamentations have ascended; and cannot cease because of the lawless deeds which are wrought upon the earth.

Classified Material

God was moved to anger at these Watchers, because of the horrible practices that they had introduced upon the Earth. And there was something else: God was angry at the fact that they had disclosed certain secrets, and were teaching them to their sons, the Nephilim. We are not told exactly what these secrets were except that they were "eternal secrets" which men of Earth were striving to learn, and which God did not intend for fallen man to discover.

Things deteriorated rapidly. And soon the Nephilim were practicing the most repulsive and revolting behavior. They began to "devour one another's flesh, and drink the blood." This cannibalism shocked even the depraved citizens of Earth, to the extent that they brought accusation against the Watchers.

Enoch is instructed by the Most High to deliver this warning to the Watchers: "Enoch, thou scribe of righteousness, go, declare to the watchers of heaven, who have left the high heaven, the holy eternal place, and have defiled themselves with women, and have done as the children of earth do, and have taken unto themselves wives: "Ye have wrought great destruction on earth: And ye shall have no peace nor forgiveness of sin."

These same Watchers in turn approached Enoch to mediate on their behalf, and to write out a petition in their favor. God, however, rejects the petition and Enoch is summoned to speak to the Watchers again: "Go, say to the Watchers of heaven, who have sent thee to intercede for them: 'Ye should intercede for men, and not men for you. Wherefore have ye left high, holy and eternal heaven, and lain with women and defiled yourselves with the daughters of men, and taken wives unto yourselves and done like the children of earth begotten giants as sons. And although ye were holy, spiritual living and eternal life, you have defiled yourselves with the blood of women, and have begotten children with the blood of flesh, have lusted after flesh and blood as those who do die and perish.'" And as for the giants or Nephilim produced by them:

> The giants, who are produced from the spirits and flesh, shall be called evil spirits upon the earth...Evil spirits have proceeded from their bodies; because they are born from men and from holy Watchers is their beginning and primal origin... And the spirits of the giants afflict, oppress, attack, do battle, and work destruction upon the earth.... And these spirits shall rise up against the children of men and against the women, because they have proceeded from them...This shall they destroy until the day of consummation, the great judgement in which the age shall be consummated over the Watchers and the godless, yea, shall be wholly consummated. And now as to the Watchers.... Say to them therefore: "Ye have no peace."

Such was the message given to Enoch, who transmitted it to the Watchers and their progeny on Earth.

The next question is: How was this information preserved for us? How did the succeeding generation learn these facts? The Book of Enoch supplies the answer: "And now, my son Methuselah, I tell thee everything and write it down for thee: I have revealed everything to thee, and handed thee the books which have to do with all these things. My son Methuselah, preserve the books that come from thy father's hand and hand them on to the coming generations of the world."

Enoch was also given a message to deliver to Noah, telling him that the whole Earth was to be destroyed. He was also to instruct Noah in the way of escape, so that his seed be preserved for all generations.

Unquestionably, the Book of Enoch confirms the biblical record that the Earth was defiled and polluted by the incursion of extraterrestrial beings, and particularly by their shameful behavior with the "daughters of men." And as in the Book of Genesis, so in the Book of Enoch, God is incensed with this sexual coupling between celestial and terrestrial beings, and begins to move in an act of terrible judgement. "And the Lord said, I will destroy man whom I have created from the face of the earth" (Genesis 6:7).

The Sons of Jared

Other Books, like *The Book of Jubilees* add a few more details to this awesome story. It reveals the date when these heavenly Watchers descended to Earth—461 Annus Mundi, a date which Bishop Usher would interpret as 3543 BC. It also notes that these Watchers were specifically associated with Jared, the fifth in line from Adam.

Interestingly, there is an American organization called "The Sons of Jared." Their publication is called, *The Jared Advocate.* The aim of this group, as far as one is able to understand it, is to declare war on all the descendants of the Watchers—including notorious kings and dictators, who have tyrannized mankind. A sort of celestial Mafia!

Regarding this association with Jared, *The Book of Jubilees* tells us: "And in the second week of the tenth jubilee of Mahalelel took unto him a wife Dinah, the daughter of Barakel, the daughter of his brother's brother and she bore him a son in the sixth year and he called his name Jared for in his days the angels of the Lord descended on the earth, those so named the Watchers."[93]

Other Ancient Documents

Another minor source of data is the *Zadokite Document.* This was discovered about a half-century ago in the attic of an old synagogue in Cairo and is thought to be related to the Dead Sea Scrolls. This document also refers to the descent of the Watchers, and to their giant offspring. "Because they walked in the stubbornness of their heart, the Watchers of heaven fell, yea, they were caught thereby because they kept not the commandments of God. So too their sons whose bodies were as mountains. They also fell." [94]

The *Apocalypse of Baruch* is another ancient document which confirms the story of the fallen angels. Written in Syriac, it adds this new concept to the story of the fallen angels: that the source of the corruption was the sinfulness of mankind. It was human sin that cause the angels to fall.

A series of booklets called *The Testament of the Twelve Patriarchs,* purporting to come from the twelve sons of Jacob,[95] also refer to the fallen angels. They contain nothing like the detail of the Book of Enoch, but they do make an interesting statement: the women of Earth were the prime movers in alluring and enticing the angels. In this they agree with the Apocalypse of Baruch. They also introduce a brand new element, suggesting that the act in which the angels and the women participated was more psychological than physical. "Thus they allured the Watchmen before the flood, for as these continually beheld them, they lusted after them and conceived the act in their mind; for they changed themselves into the shape of men and appeared to them when they were with their husbands; and the women, lusting in their minds after their forms, gave birth to giants."[96] It appears from this Testament of the Twelve Patriarchs that the "lusting" and the "conceiving" *were in the mind.* Nonetheless, the results were physical enough—they gave birth to giants!

Jewish documents are not the only ones which refer to fallen angels. The Koran contains a brief reference to two angels, Harut and Marut, who came to Earth. Their fall takes place when a beautiful woman appears before them. These angels had been warned in advance that there were three things that they were not to do: admit the existence of other gods; murder; drink intoxicants. It was precisely these three things that the charming woman demanded, if the angels were to enjoy her favors. Finally, the angels agreed to drink wine—as the least of the three sins—and their paramour granted their desires.

Folklore and Fables

Another source of information is the abundance of myths, legends, folklore and fables that speak of "giants" upon the Earth in ancient times, and how there was sexual union between demigods from

Heaven and women from Earth. Many scholars believe that myths emerged from a kernel of historical fact. According to Andrew Thomas, mythology and folklore are "thought-fossils depicting the story of vanished cultures in symbols and allegories."[97] If this is so, do we find anything in myths and legends resembling the bizarre event of Genesis 6? There are indeed numerous such traditions among many nations.

Most people are aquatinted with the mythologies of ancient Greece and Rome. The gods or semi-gods in these traditions go under different names, but their behavior has a common denominator. Whether these gods are called Zeus or Jupiter, Poseidon or Neptune, Aphrodite or Venus, Eros or Cupid…their sex orgies, promiscuities, cruelties and violence are all of the same cloth. And so are their offspring.

The Genesis story, according to Tom Horner, corresponds precisely to the Age of the Heroes in Ancient Greece. These heroes were also "spawned by divine fathers and human mothers. One of them was Hercules."[98]

The story of Zeus also is well known. Promotheus was aware of the secret that Zeus had no control over his lusts, and aware also of the names of women whom he would seduce. Because if this and other reasons, Zeus planned for Promotheus to be chained in the Caucasius, where an eagle would feed on his liver each day. But each night his liver would be renewed. In this way. The torture of Promotheus was endless. Eventually however, he and Zeus were reconciled. But cruelty was not the only distinctive of Zeus. There seemed to be no boundaries or limits to his lust, and numerous women were seduced by him, including Thetis, Europa, Leda, Metis, and Dione. Emile Gaverluk says: "Zeus' amorous victories illustrate the actions of uncontrolled spirit-beings lusting after human flesh. **The whole story of Greek mythology is an expanded version of that astonishing verse in the Bible: 'The sons of God saw the daughters of men that they were fair; and took them wives of all they chose'** (Genesis 6:2)…The mythology of the past is a startling revelation of the uncontrolled behavior of both spirit-beings and rebellious man."[99]

But the mythologies of Greece and Rome are not the only ones that relate such strange events. Erick Von Däniken has supplied us with a wealth of samples from around the world.[100] We admit that his speculative outbursts frequently outmatch his theological soundness; nevertheless, we are indebted to him for bringing to our attention many incontrovertible facts which others had left buried and undisturbed.

> Ancient Sumerian records tell of gods descending from the stars and fertilizing their ancestors. This interbreeding of gods from heaven and women from earth is supposed to have produced the first men upon the earth.
>
> The native inhabitants of Malekula, in the New Hebrides believe that the first race of men were direct descendants of the sons of heaven.
>
> The Incas held that they were the descendants of the "sons of the Sun."
>
> The Teutons claimed that their ancestors came with the flying **Wanen.**
>
> Some of the South Sea Islanders trace their ancestry to one of the gods of heaven, who visited them in an enormous gleaming egg.
>
> The Koreans believed that a heavenly king, "Hwanin," sent his son, "Hwanung," to earth, married an earth woman who gave birth to Tangun Waggom. It was he who was supposed to have welded all the primitive tribes together into one kingdom.
>
> The ancient tradition Tango-Fudoki in Japan tells the story of the Island Child. The only difference here is that it was a man from earth and a maiden from heaven that came together in marriage, and spent their time together in heaven and not on earth.
>
> From India comes the Mahabharata and other ancient

> Sanskrit texts, which tell of "gods" begetting children with women of earth, and how these children inherited the "supernatural" skills and learning of their fathers.
>
> A similar mythology is found in the Epic of Gilgamesh, where we read of "watchers" from outer space coming to planet Earth, and producing giants.
>
> An early Persian myth tells that before the coming of Zoroaster, demons had corrupted the Earth, and allied themselves with women.

When these and many other accounts, are all tied together, they amaze us by their common core. Each one refers, with slight variations, to the traffic between "sons of God" and the "daughters of men," to the sexual activities in which they engaged, and to the unusual and abnormal offspring they produced. A further convincing element in this string of samples, is that these myths and legends belong to people so far removed from each other by time, space, and language that collaboration or conspiracy is out of the question. How then does one explain this phenomenon?

Fable or Fact?

Could it be, that at some distant point in time, these bizarre events actually did take place? Rather than being the fertile product of the imagination of primitive man, they were actually man's crude description of what actually happened? Beings from Heaven and beings from Earth did actually come together, did generate children, and did produce the Nephilim.

Later, of course, many of these myths and legends developed an overgrowth of fictional imagery and imaginative fantasy, which clung to them like moss to the mill. There can be little doubt, however, that beneath this dense overgrowth lies the plant of truth. What began as history, ended up as legend.

Above all else, we are convinced that the Bible speaks of these things. The basic message of the apocryphal documents and the various mythologies find corroboration in God's word. Spirit-beings from outer space did arrive on Earth, and did produce these unnatural offspring. By this cross-breeding, man became more and more absorbed with evil and violence; the human race became tainted and corrupted; and soon the Earth was unsalvageable. These extraterrestrial beings by means of genetic manipulation had succeeded in contaminating God's special creation; so much so, that their destruction, as well as that of man, became a moral necessity.

7

Descent into Hell

Calvary must have had an immediate
and tremendous effect upon that spirit-world,
the full extent and nature of which we may not yet know.

—W. Graham Scroggie

One would hardly expect the Apostles' Creed, composed in the early Christian era, to refer to the Nephilim. Such a document seems far removed from the events of Genesis 6. But let's take a look.

The Apostles' Creed is a distillation of doctrine, abbreviated down to an "irreducible minimum." Because a creed demands such condensation many a truth has to be omitted, and only major, cardinal truths are included. In the Apostles' Creed, truths about our Lord's teaching, preaching, miracles…Have been omitted, and so has all reference to the events of Pentecost. Not that these were unimportant; it was that the creed formulators had to be fastidiously selective. A truth had to be absolutely paramount to gain admittance into this Creed.

Where does this lead us? To the all important question of our study: Does the Apostles' Creed contain any reference to Genesis 6, and to the Nephilim? It certainly does. On the surface, it may not be all that apparent, but it is there.

Embedded in the Creed is an article which received but scant attention from modern preachers and professors. It could well be called "the forgotten article." It reads: "*He descended into hell.*" In the three-day interval between His death and resurrection, Christ went to Hell to fulfill a specific mission.

Admittedly, this particular article does not appear in the earlier forms of the Apostles' Creed. However, it was an integral art of the Apostles' beliefs, and was later included in the Creed itself. The point we are making is, that the inclusion of this particular article, "He descended into Hell," had to be of special significance indeed. As special as that of the Virgin Birth, the Cross, the Resurrection, the Second Advent...

It is thus all the more surprising that this truth is rarely discussed anymore; rarer still is it preached; and still rarer does anyone sing about it. It was not always so. In the 1641 Prayer Book there is included the Sternhold and Hopkins version of the Psalms, and in it we read:

His soul did after this descend
Into the lower parts;
A dread unto the wicked spirits,
But joy to faithful hearts.

The biblical passage that sheds most light on this article is 1 Peter 3:18–20. "For Christ also hath once suffered for sins, the just for the unjust, that he might bring us to God, being put to death in the flesh, but quickened by the Spirit: By which also he went and preached unto the spirits in prison; Which sometime were disobedient, when once the long-suffering of God waited in the days of Noah, while the ark was a preparing, wherein few, that is, eight souls were saved by water." Interestingly, this is the passage designated in the Prayer Book to be read on Easter Eve, coinciding perfectly with biblical chronology. And for the morning of the same day the designated reading was Zechariah 9, which speaks of the "pit wherein is no water" and of the "prisoners of hope."

Interpretations

Numerous and varied attempts have been made to interpret this passage from Peter. It is doubtful if any passage in the Bible has had more interpretations.

- Some have interpreted it as a warrant for purgatory.
- Some as a warrant for universalism, claiming release for prisoners of all time.
- Some as a warrant for the release of those prisoners who were in Hell at that particular time, and referred to as the "harrowing of hell."
- Some as a probationary period for immature saints.

But all of these theories run contrary to the testimony of Scripture. Nowhere are we taught that the saints have to go through a process of purging and purifying before gaining admission to Heaven. Man enters Heaven not by the slow, purifying process of purgatory which he endures, but by being made acceptable to God in the perfection of Christ who endured on his behalf. To argue universalism on the basis of this text is not only to deny the testimony of other passages of Scripture (John 8:24, Hebrews 9:27) but it is to deny the purpose of Christ in coming to the world.

Obviously, these theories do not interpret the text. What, then, is its meaning? What does the Bible mean when it declares that Christ descended to preach "unto the spirits in prison?" (1 Peter 3:19). Here again we are presented with a variety of interpretations, even among Evangelicals.

Calvin dismisses the passage with a general comment that Christ descended into Hell in order to complete His vicarious suffering. By going there He endured for a few brief hours the torments of the lost. It was a part of what Calvin called the "horribiles angustiae." But this interpretation sheds no light on the reference to Noah, nor on why should Christ preach in Hell.

- Some interpret the passage as meaning that Christ went to preach to all the Old Testament dead; and that those who were disobedient in Noah's day were representative of all others. But the passage simply does not say this.
- Some suggest that Christ actually preached in the days of Noah to the spirits who were afterwards confined to prison. Again, conjecture.
- Augustine's theory was somewhat similar. He stated that Christ preached **in the spirit** in the days of Noah, Just as He preached in the flesh in the days of Galilee. But if such were the case, it still does not shed light on the descent into Hell.
- In the same vein others theorize that the reference has nothing at all to do with the Lord's action between His death and resurrection. Nor, as a matter of fact, does it refer to the days of Noah. Rather it refers to the preaching of the apostles after the resurrection of Christ. Frankly, we fail to see the connection.

It is obvious that all these so-called "explanations" leave a lot of questions to be answered. Who are these spirits in prison? Why were they there? Why specifically those of Noah's day? What exactly did Christ preach to them?

Peter's Motives

Before beginning to answer each question individually, it is essential to know what the motives of Peter in writing this letter were. We know that he refers to Christ preaching unto the spirits in prison, but how does this fit in with the total message of the letter?

Fortunately, Peter's message and motive is crystal clear. It is to encourage the saints who were suffering persecution for their faith. The letter abounds with references to them. They are the elect of God "scattered" by a terrible persecution unleashed upon them by the Roman Emperor. They had become targeted for special treatment during the reign of Nero in AD 64–68. It proved to be one of the fiercest and cruelest persecutions in the history of the church.

Peter is concerned about the saints who had been dispersed and scattered throughout the empire, and writes this letter to encourage their hearts and strengthen their faith. "That the trial of your faith, being much more precious than of gold that perisheth, though it be tried with fire, might be found unto praise and honor and glory at the appearing of Jesus Christ" (1 Peter 1:7). "Beloved, think it not strange the fiery trial which is to try you, as though some strange thing happened unto you: But rejoice, in as much as ye are partakers of Christ's suffering; that, when His glory shall be revealed, ye may be glad also with exceeding joy" (1 Peter 4:12–13).

Reign of Terror

There is no doubt about the "fiery" element. This is exactly what awaited some of the saints at the hands of Nero. He used them as flaming torches to illuminate his gardens at night— referred to by Roman historians as "flammati."

Tacticus records: "Their sufferings at the execution were aggravated by the insult and mockery: for some were disguised in the skins of the wild beasts and worried by dogs; some were crucified; and others were wrapped in pitched shirts and set on fire when the day closed, that they might serve as lights to illuminate the night."[101]

Juvenal writes in the same vein: "Burning in their own flame and smoke, their head being held up by a stake fixed to their chin till they made a long stream (of blood and sulphur) on the ground.[102]

Peter underscores the same message in chapter 3 of his letter. He tells then not to be afraid of this terror, but to rejoice because they are suffering for righteousness sake. After all, it is better to suffer for well-doing than for evil-doing (1 Peter 3:17). Then he reminds them of what happened to the Savior Himself. He, too, suffered, "the just for the unjust." And then comes the thrust of his whole message: *note what happened to Jesus afterwards.* After He suffered and was put to death in the flesh, He was quickened by the Spirit (King James) or in the Spirit (New American Standard) (1 Peter 3:18). He exchanged the "flesh" by which alone He could suffer, for a glorious, spiritual body which could neither suffer nor die. Then came the descent into hell, followed by the resurrection and then the ascension. Christ is now "gone into heaven, and is on the right hand of God…" (1 Peter 3:22). The same thing will happen to His followers who are suffering and even being martyred for their faith. They, too, have been begotten "unto a lively hope by the resurrection of Jesus Christ from the dead, to an inheritance incorruptible, and undefiled, and that fadeth not away, reserved in heaven for you…(1 Peter 1:3–4).

This is the whole purpose for Peter's letter: to comfort and encourage the persecuted saints by depicting the glory to come. There is a resurrection awaiting them. Their death will issue in victory and triumph just as it did in the case of Jesus. As for Jesus, he tells them, His triumph was such that he went and announced it to "the spirits in prison."

To understand what exactly is meant by this statement "preaching unto the spirits in prison," it is necessary to examine the three major words used by Peter in this text: "preach," "spirits," "prison."

Preach

We normally associate preaching with proclaiming the gospel. But actually this is to put a limitation on the word which it does not have in the original. The Greek word for "preach" here is *kerusso,* and means to herald, to publish, to proclaim. It does not tell us what it is to be heralded, published, or proclaimed. There is another Greek word that does that, and it is *evangelizo.* That word means specifically to preach the gospel of preach the Good News. But *kerusso* is different; by itself it tells nothing of the message to be proclaimed.

Kenneth S. Wuest in his *Word Studies* tells us that the word (kerusso) in itself gives no indication of the content of the message. A qualifying phrase must be added for that purpose.

This is why when the word *kerusso* is used in the New Testament there is usually added to it a word like "gospel" or "God." "Jesus went about all Galilee, teaching in their synagogues and preaching the gospel of the Kingdom…" (Matthew 4:23). "We preached unto you the gospel of God" (1 Thessalonians 2:9). The bottom line is: when the word *kerusso* is used alone as in 1 Peter 3:19 without any reference to the terms of the proclamation then in no sense is the gospel or salvation a necessary part of its meaning. Thus when Jesus went and preached unto the spirits in prison, it did not necessarily mean that He went to preach the gospel to them. It simply meant that He went and *announced something* to them. What that something was, will become evident as we proceed. It is sufficient to say at this juncture that it was not the gospel but something of a judgmental nature. Anyway, angels were never included among those for whom the Lord died. Preaching the gospel has no relevance for a congregation of angels.

Spirits

Just as the word *preach* has been invariably associated in our thinking with preaching the gospel, so the word *spirits* has been invariably associated with the spirits of *men.* But again this is to assume an association that does not exist. In the original, when the word *spirits* stands alone it never signifies men. When it does signify men there is always added to it a further definition, like a qualifying word or clause, e.g., "A certain damsel possessed with the spirit of divination" (Acts 16:16). "The spirits of just men made perfect" (Hebrews 12:23). The word spirits (pneumata) by itself, without any qualifying description, is always used of *supernatural beings,* higher than man and lower than God. Beings that have no corporeal garb of "flesh and blood" or "flesh and bones."

Used without a qualifying addition, spirits mean supernatural beings. E.g., "Notwithstanding, in this rejoice not, that the spirits are subject unto you; but rather rejoice, because your names are written in heaven" (Luke 10:20). Spirits here obviously mean demons. "When the even was come, they brought unto Him many that were possessed with devils: and He cast out the spirits with His word, and healed all that were sick" (Matthew 8:16). "Now the Spirit speaketh and expressly that in the latter times some shall depart from the faith, giving heed to seducing spirits, and doctrines of devils" (1 Timothy 4:1).

It is clear that when the word *spirits* is used alone it invariably refers to supernatural beings. Sometimes good angels, sometimes fallen angels. There is a specific reference in Psalm 104:4: "who maketh His angels spirits." And the same truth with even greater clarity is found in Acts 8:26–39. In verse 26 God's messenger is referred to as "the angel of the Lord"; but in verse 29 as "the spirit." And in verse 39 again as "the spirit of the Lord." The words *angel* and *spirit* are obviously used interchangeably. Revelation 1:4 speaks of the seven spirits that are before the throne of God; and Revelation 4:5 of the "seven spirits of God." These were seven angelic beings standing in a special relationship with the Lord. Even more conclusive is the fact that the two words come together in Acts 23:8. The Sadducees, we are told, believe in "neither angel nor spirit."[103]

Let one more Scripture suffice. Paul in a famous passage in 1 Timothy 3:16 writes: "And without controversy great is the mystery of godliness: God was manifest in the flesh, justified in the Spirit, *seen of angels…*" (emphasis added). A superior translation would be: "beheld by angels" (New American Standard) or "seen by angels" (New International Version). But when was He seen by angels? During the three days when He preached to the spirits in prison.

In answer to the question "Who are these spirits?" Kenneth S.

Wuest says: "They cannot be human beings...The word *pneuma* is used as a designation of just two classes of free moral agents in the New Testament, of angels (Hebrews 1:7, 14) and of demons (Matthew 8:16, Luke 10:17, 20).[104]

It is interesting to observe how the apostles complement each other. The Apostle Paul tells us that Christ appeared to angels (1 Timothy 3:16); the Apostle Peter, why and where Christ appeared to angels (1 Peter 3:19; 2 Peter 2:4); and the Apostle Jude why the angels were there in the first place (Jude 6, 7).

It is obvious that not all fallen angels have been incarcerated to prison, or else there would be no demons free today to afflict the human race. God incarcerated only those angels who had been "disobedient in the days of Noah," i.e., those angels who had sinned with the women of Earth and by so doing had corrupted and tainted the human seed. In other words, the clue to the meaning of the article in the Apostles' Creed is 1 Peter 3:19, and the clue to 1 Peter 3:19 is Genesis 6.

Prison

One author has drawn attention to the fact that the word prison received emphatic prominence in the original structure of this verse. Coming at the end it signifies: *"even to the spirits in prison,"* as if it were a news event of unusual significance that Christ should have made this proclamation in the prison itself. Even there.[105]

Do we have any information as to the location of the prison? Peter in 2 Peter 2:4 uses the Greek word *Tartarus*—the only time it is used in Scripture. And used, let it be noted, with reference to fallen angels and not fallen men. The word has been variously translated in our English versions: the NEB gives us "the dark pits of hell"; the RV "pits of darkness"; the NIV "gloomy dungeons." One translator renders it, "the vilest province of hell." In all these translations, depth and darkness are the distinctives of Tartarus.

When Homer used the word he gave it the meaning of *subterranean.* Hades was the place where the souls of departed men awaited the coming judgement, but Tartarus was a much deeper and darker abyss and reserved specifically for fallen angels. One is reminded of the lines of Milton: "And in the lowest deep, / Still threatening to devour me opens wide."[106]

Similarly, the Book of Enoch (22.2) reserves Tartarus for these same fallen angels.

In view of all this, how does one interpret 1 Peter 3:19? It is obvious that we cannot adopt the popular but erroneous interpretation declaring Christ went to Hades to preach the gospel to the spirits of men who had sinned in the days of Noah, intending to extend them another chance. This view in no way fits the purpose of the letter, which was to encourage the saints to endure persecution for the Lord's sake. Bullinger states it eloquently:

> Note the incongruity and inconsequence of the popular explanation, which is to this effect: "Christ also suffered, and after He died, He went and preached the Gospel to the greatest evil-doers the world has ever seen, so great that their sins brought down the judgement of the Flood!" We ask, What has this to do with the argument of the Holy Spirit in the context? What **reason** is this, why is it good to suffer for the Lord's sake? What encouragement is there in this for them, or for us to suffer for well doing? Apart from the inexplicable supposition that these greatest of sinners are singled out for special mercy, this interpretation is really at variance with the argument! It would be, indeed, rather an argument for evil doing rather than for well doing! For why should we suffer for well doing when, even if we do evil, Christ Himself gives us hope of salvation after death.[107]

It seems clear that such an interpretation of 1 Peter 3:19 does injustice to the meaning of the text, is contrary to the purpose of the letter, and is unscriptural as regards the doctrine it purposes.

The Triumphant Descent

The only interruption that does justice to all three categories—the meaning of the text, the purpose of the letter and the doctrine of Scripture—is the one that describes Christ going down to the deepest dungeons of Hell, not to preach salvation to lost men but rather to proclaim His victory to the fallen angels. That victory had been completed at the Cross, and thus the judgement of the angels had been sealed. So complete was that victory that Christ went down even to Tartarus to announce it. So complete was His victory that "having spoiled principalities and powers, He made a shew of them, openly, triumphing over them in it" (Colossians 2:15). So complete that "angels and authorities and powers" have been made "subject unto Him." So complete that things in Heaven, things in Earth, and even things under the Earth, may know for evermore that He is Lord of all.

This view has the additional merit of keeping in sequence the whole series of Christ's actions between His death and His ascension. The chronological order is perfect: "put to death in the flesh," "went," "preached," "resurrected," "gone into heaven" (1 Peter 3:18–22).

But to return to our original quest, the position is this: down in the dark dungeons of Hell are certain fallen angels, incarcerated there for a sin they committed back in the days before the Flood. Their sin was specifically that of leaving their own habitation; giving themselves over to fornication; and going after strange flesh. In other words, these extraterrestrial beings had lusted after carnal relations with inhabitants of Earth.

Demonic Inbreeding

This specific sin of which these extraterrestrials were guilty, must be classified as the most repugnant and repulsive that one can meet anywhere in the annals of the human race. It is such an abomination in the sight of the Lord that it called forth the ultimate judgement. No sin ever moved the Sovereign Lord of the universe to such unmatched anger and unmitigated wrath. And no sin was ever punished like this one. Its magnitude was such that both parties suffered in the most unusual and devastating manner—humans by being nearly exterminated in the Flood and angels by being committed to "everlasting chains." And what is of frightening import to us is that this same sin may well be committed again in the last days of Planet Earth. "And as it was in the days of Noah, so shall it be also in the days of the Son of man" (Luke 17:26).

But however black this evil was in itself, it was also indicative of an even blacker transgression. It signified nothing less than the total disarray and derangement of the Divine Plan for the redemption of the human race. If Satan had succeeded in his conspiracy, it would have made the Incarnation of Jesus and the Atonement of Jesus impossible. If the human race had become a hybrid mixture of angelic and human, then Christ would not have been able to answer in His humanity to the humanity of man. What is more, God had no plan, and has no plan, for the redemption of angels (Hebrews 2:16). God's plan was for "the seed of Abraham." God by-passed the fallen angels, leaving them chained in Tartarus, but in infinite mercy He did plan the salvation of fallen man through the vicarious sacrifice of His own Son. But this whole plan would have been in jeopardy if the human race had become permanently polluted by the inbreeding of demonic agencies from Hell.

8

Who Are the Nephilim?

. . .Climax at the beginning.

—Arthur Custance

Of all the imaginable phenomenon Earth, the progeny of this union between extraterrestrials and humans is the most bizarre. Man has paid little attention to them until now, for the fact of their little existence has been shrouded in legend. But can they be dismissed as myth any longer? In this end-time hour many strange phenomena are occurring. Jesus said, "As the days of Noah were, so shall also the coming of the Son of man be" (Matthew 24:37). It was the wickedness of humanity and the abominable union of the supernatural with the natural that moved God to judge the world. We are fast approaching a new period of God's wrath. The return of these super creatures may even now be a threat to us.

The Nephilim: Etymological Evidence

Who are these beings? A clue to their identity is found in their name—Nephilim. The word itself is Hebrew and is first used in Genesis 6:4. "There were giants in the earth in those days: and also after that, when the sons of God came in unto the daughters of men, and they bare children to them, the same became mighty men of which were of old, men of renown." Nephilim is translated "giants" in the Authorized King James Version, but "giants" is in no way a complete description.

Commentators like Lange trace the word "Nephilim" to the root "Niphal meaning "Distinguished ones." This corresponds perfectly with the "men of renown" at the end of Genesis 6:4, nevertheless it is not a generally accepted translation.

Others have sought the root of the word in the Hebrew consonants "npl" as found in Psalm 58:8. Here it means "miscarriage." Accepting this theory, the Nephilim would be those superhuman beings that resulted from miscarriages. Genesis Rabbah (26:7) seems to confirm this translation when it states: "Nephilim denotes that they hurled the word down, themselves fell (naflu) from the world, and filled the world with abortions (Nephilim) through their immorality."[108]

Most scholars, however, reject both these interpretations and trace the word "Nephilim" to the Hebrew root "Naphal" meaning "to fall." The Nephilim are the "fallen ones." A direct reference to the fallen angels who sired them.[109]

Because of some uncertainty in the translation of the Hebrew word, more and more Bible versions are now leaving the original word untranslated. Thus the New International Version renders the passage: "The Nephilim were on the earth in those days…" (Genesis 6:4). This also does justice to the fact that the definite article precedes the word in the original. "The Nephilim were on the earth…" (emphasis added). The same definite article is also found in the other biblical passage where the word "Nephilim" occurs, namely, Numbers 13:33. "We saw the Nephilim there…" (emphasis added).

Interpreting the Nephilim as "the fallen ones" dismisses the suggestion of one modern author that belief in the Nephilim could lead to racism. He fears that people would boast of having celestial blood in their veins and consider themselves superior to those of the ordinary, human ancestry. To claim descent from stellar explorers could lead to a class distinction surpassing anything ever seen in society, but he need have no such qualms! If the Nephilim are stellar, they are also infernal. To claim descent from such beings would be to admit an ancestry from Hell.

Although there is no etymological evidence to justify "giants" as an accurate translation for "Nephilim," such a translation is not without merit. In more senses than one, Nephilim were giants.

Giants in Size

For one thing, they were giants in *size* and *strength*. Much documentation of the exceptional physical stature and superhuman strength of the Nephilim exists, and this is not in the least surprising, knowing that they were "fathered" by angels. Angels as already stated, do "excel in strength" (Psalm 103:20). They are greater "in power and might" than men of earth (2 Peter 2:11).

What applies to holy angels, applies equally to rebellious angels. Their moral fall does not seem to have reduced their physical prowess. The Book of Enoch states that their "height was like the tallness of cedars and whose bodies were like mountains."

A modern author describes them in the following terms: "Perverted power and strength are (thus) conspicuous attributes of fallen angels. This titanic energy is displayed in the supernatural strength demons can impart to the human body when they enter and possess it."[110]

The New Testament supplies many such examples. One of the most noted is that of the Gaderene demoniac, who by his superhuman strength could snap fetters and break chains.

The Roman Catholic Church confirms this attribute of fallen angels when it demands the presence of a superhuman strength before it will diagnose a person as being demon-possessed. Actually, a person under investigation must reveal the presence of *three* phenomena before the Roman Catholic Church will categorize him as "possessed." First, he must be able to speak in a language unknown to him. Second, he must have knowledge of secret facts, previously unknown to him. Third, he must possess unnatural strength beyond his age and ability.[111]

Dr. Kurt Koch, from his vast experience and extensive research into occultism, has discovered that even children or delicately built women can offer effective resistance to three or four strong men when demon possessed. [112]

Similarly, Professor Oesterreich cites a number of examples from his research, demonstrating the same superhuman strength in demon-possessed people. One example he gives is of a ten-year-old boy who could scarcely be held down by three adults. Another is of a young girl who could barely be controlled by two men.[113]

In a similar vein, Robert Pearson wrote from Borneo in 1967 concerning the Dyak uprising:

> Evidences of demonic power were witnessed at Andjungan. Dyaks used their fists and feet to break display cases with glass flying all over the place. Some actually danced on it with bare feet but no one was injured. One missionary watched Dyaks step into pans of acid used to coagulate rubber. Undiluted, this acid can normally burn the flesh to the bone, but these men were unharmed. Others struck locked and barred doors with their bare hands, breaking them down so easily as if they had been rammed by a truck...These things are hard to understand, but we know that Satan is powerful and...able to endow men with his power when it suits his purpose.[114]

John Wesley in his journal entry for May 2, 1739, writes of a certain John Haydon, a respectable person, present in one of his meetings, and who "fell off his chair and began screaming terribly and beating himself against the ground…two or three men were holding him as well as they could. The man then roared out, 'O thou devil! Thou cursed devil! Yes thou legion of devils! Thou canst not stay! Christ will cast thee out!' Fortunately, after prayer, Wesley reports that "both his body and soul were set at liberty."

These and countless similar examples are reminiscent of what happened to the seven sons of Sceva in Acts 19. A demonized man out-numbered seven to one was able to overcome all seven by a phenomenal show of strength. The Bible states that they were fortunate to escape from the house naked and wounded.

A Second Incursion

In Genesis 6, where the word "Nephilim" is first used, we are told that the Nephilim appeared on the Earth just before the Flood, and that their appearance was the main reason for the Flood. There followed another incursion of these fallen angels at a later date. We read in Genesis 6:4, "The Nephilim were on the earth in those days—and also afterward…" (NIV). This data is found in Numbers 13:33: "We saw the Nephilim there (the descendants of Anak come from the Nephilim)" (NIV).

This second eruption was probably on a more limited and restricted scale than the first. Nevertheless, God ordered their complete destruction.

One has often wondered at the severe and extreme measures that God asked Joshua to put into effect once he entered the land of Canaan. God commanded him to "utterly destroy them." We have found it difficult to reconcile this with the character of God. One can only surmise that God had a special reason to issue such a command. Could it be that God knew the heritage of these Nephilim?

God was aware that the Canaanites and their neighbors manifested the whole gamut of demonical practices, and that they were a threat to the character and destiny of His chosen people, who were entering the land at that time. This is why He warned the Israelites not to imitate the occult practices of these people whom they dispossessed. With the same unmatched anger that He had displayed in Genesis 6, God orders the complete extermination of the inhabitants of Canaan. "But thou shalt utterly destroy them; namely, the Hittites, the Amorites, the Canaanites, and the Perizzites, the Hivites, and the Jebusites; as the Lord thy God hath commanded thee" (Deuteronomy 20:17). However, Israel, as so often in her history, failed to obey God, and there is reason to believe that some of the Nephilim survived (Joshua 13:13, 16:10; Judges 1:28–34).

The progeny of these Nephilim went under various names. We read of the Anakim, descended from Rapha; the Zamzummims, the Emims, the Avims, etc. All shared the characteristics of being huge, tall, and strong. Rabbi Bahya ben Asher, a Spanish Cabalist, claimed the Nephilim were heads of the family called "sons of God." They were so called because terror fell on those who saw them. As the virility of the stock declined, they were called Anakim and later Rephaim.[115] Here is an Old Testament description of the Emim: "The Emim dwelt therein in times past, a people, great and many, and tall, as the Anakim; which also were accounted giants, as the Anakim; but the Moabites called them Emims" (Deuteronomy 2:10,11).

These men were such giants that the Israelite spies who went in to reconnoiter the land, cowered before them. "And there we saw the giants, the sons of Anak, which come of the giants; and we were in our own sight as grasshoppers, and so we were in their sight" (Numbers 13:33).

Flavius Josephus, the noted Jewish historian of the first century AD, described these giants as having "bodies so large and countenances so entirely different from other men that they were surprising to the sight and terrible to the hearing"[116] And he adds that in his day, the bones of the giants were still on display!

Deuteronomy 3:11 describes one of these giants in more detail:

"For only Og king of Bashan remained of the remnant of giants: behold, his bedstead was a bedstead of iron; is it not in Rabbath of the children of Ammon? Nine cubits was the length thereof, and four cubits the breadth of it, after the cubit of a man." A super king-sized bed! In modern measurements it was 18 feet, 6 inches long, and 8 feet, 4 inches wide!

Some of these giants carried spears that weighed from ten to twenty-five pounds. One carried a spear whose staff was "like a weaver's beam" (2 Samuel 21:19). Goliath wore a coat of armor that weighed 196 pounds, and he was said to be about nine feet tall. Some of these giants had six fingers on each hand and six toes on each foot.

Evidence from Outside the Middle East

These giants were not confined to the Middle East. Two dozen human footprints of abnormal size have been found in the Paluxi riverbed. Texas, some of them measuring eighteen inches long. Other giant markings have been discovered in such diverse places as Colorado, New Mexico, Arizona and California. In the Mt. Vernon area of Ohio, Dr. Wilbur G. Burroughs of the Geological Division of the Berea College, Kentucky, reported finds of human foot prints 23.75 cm. long and 10.25 wide! Near Antelope Springs, Utah, William Meister discovered in 1968 two human footprints 32.5 cm. Long and 11.25 wide.

Similar giant footprints have been discovered in other countries especially in the Mt. Victoria region of Australia.

Not only do we have footprints of giants but actual skeletons as well. In 1936 Larson Kohl, the German paleontologist and anthropologist, found the bones of gigantic men on the shore of Lake Elyasi in Central Africa. Other giant skeletons were later found in Hava, the Transvaal and China. The evidence for the existence of giants is incontrovertible. "A scientifically assured fact," says Dr. Louis Burkhalter.[117]

Giants in Knowledge

The Nephilim also were giants in *knowledge.* According to the Book of Enoch, God was incensed against the fallen angels partly because they disclosed certain classified information to humans. The ancient world associated demons with special knowledge and with superior intelligence. The word "demon" in Greek (*daimon*) comes from the root meaning "knowledge" or "intelligence." The Scriptures also testify to the fact that demons have access to knowledge and information denied to ordinary mortals. We read in the Gospels how they recognized and acknowledged the deity of Christ when humans seemed totally blind to the fact. When the Gaderene demoniac saw Jesus, he fell down before him and cried out, "What have I to do with thee, Jesus, thou Son of God most high?" (Luke 8:28). These demons recognized Jesus at the beginning of His ministry, way ahead of His own disciples.

In the Book of Acts, with the same supernatural knowledge, demons recognized the mission and message of the Apostle Paul. The possessed damsel at Philippi cried, "These men are the servants of the most high God, which shew unto us the way of salvation" (Acts 16:17). This happened at a time when the people of Philippi had no idea who Paul was, nor did they know the nature of his mission. We cannot but note that every recorded statement made by demons in the New Testament concerning Christ or Paul was 100 percent accurate.

Clement of Alexandria suggested an interesting reason for this superior knowledge of demons: "It is evident, since they are demoniac spirits, that they know some things more quickly and more perfectly than men, for they are not retarded in learning by the heaviness of a body."

Examples of this trait in demons is supplied by various missionary organizations. They tell us how people possessed by evil spirits acquire superior knowledge, far above that of their brethren. In an editorial *Christian Life* Magazine we find these words concerning the murder of five young American missionaries in the jungles of Ecuador in 1956:

> Indians at Arajuno mission base knew in a few hours what had happened when five missionaries deep in Ecuador's Auca territory in 1956 failed to make radio contact with anxiously waiting fellow missionaries. How? They asked a local witch doctor. He obliged by falling into a trance, calling up his favorite demons and asking them to tell him where the missing missionaries were. According to the friendly Indians, they heard demons leave the scene, and in a short time, returned with the message that the missionaries were in the Curaray River with Auca lances in them.

John L. Nevius, a medical missionary in China at the end of the last century gathered a significant compendium of data on this subject. After sending a detailed questionnaire to Protestant missionaries all over China, he gleaned a vast amount of information about the symptoms of demonism. One characteristic he found again and again was the prevalence of superior knowledge and intelligence in the possessed person—even on subject matters of which the person had no prior knowledge.

Programmed from Space

This may well supply the key to the great knowledge and expertise that characterized certain men in ancient times. As already seen. Such knowledge and expertise by "primitive" men continue to defy explanation. P. J. Wiseman admits to this mysterious factor: "It was expected that the more ancient the period, the more primitive would excavators find it to be, until traces of civilization ceased altogether and aboriginal man appeared. Neither in Babylonia, nor Egypt, the land of the oldest known habitations of man, has this been the case."[118]

Arthur Custance pertinently states this strange sequence right at the beginning of human history: "...An unbelievably long time with almost no growth; a sudden spurt leading within a very few centuries to a remarkably high culture; a gradual slowing up, and decay, followed only much later by recovery if lost arts and by development of new ones leading ultimately to the creation of our modern world. What was the agency which operated for that short period of time to so greatly accelerate the process of cultural development and produce such remarkable results?"[119]

Could this agency be the Nephilim? Could this expertise have been imparted by beings from outer space? And could this explain the mysteries surrounding Stonehenge, the Mayan Caracol, Tiahuanaco, the Bay of Pisco, and particularly the Great Pyramid? Did the knowledge necessary to construct these monolith structures come from the Nephilim? Were they the ones responsible for what Custance calls the "climax at the beginning"?[120]

As for the Great Pyramid, many scientists suggest a date back in generations preceding the Flood. If so, those happen to be the very times of the Nephilim, the generations of the "giants" and of the "men of renown." But what if it could be proven that the Great Pyramid was not built until *after* the Flood? There is still no problem.

Could not Noah and his family have carried this information with them into the Ark, and transmitted it later to their descendants? According to the Babylonian version of the Flood, the "Chaldean" Noah was made to bury his books before the Flood, and then disentomb them after emerging from the Ark. But there is still another possibility: Could this knowledge have come from the second eruption of the Nephilim, which occurred *after* the Flood?

"Tree of Knowledge"

If we go further back, could one find a linkage between this esoteric knowledge and the "tree of knowledge" found in the Garden of Eden? We know that this was the one tree whose fruit Adam and Eve were forbidden to eat, or even touch (Genesis 2:17, 3:3). Why was this tree so named? Few commentators have shed any light on its meaning. Many dismiss the account as mere symbolism, without even telling us what the symbolism is supposed to represent. Others look upon the passage as poetry, conveniently forgetting poetry is often the truest history. It not only relates facts but interprets them as well. Poetry has been described as "history written from the inside rather than from the outside, and therefore incomparably truer."[121] Or in the words of British author Graham Green, "Poetry is the photography of the invisible."

Supposing we applied this principle to the "tree of knowledge." If it is a symbol, then obviously it must be a symbol of something. Or if poetry, then if must convey some inner truth. But why should one speculate and search for cryptic meanings when the truth may well be on the surface. Obviously, the "tree of *knowledge*" must have something to do with knowledge, or why should it be so named? It undoubtedly contained the key to certain divinely classified material that God did not want early man to possess. But somehow, and it could well be by means of the Nephilim, early man did come to possess that knowledge—at least a part of it.

This was knowledge that primitive man could never have discovered on his own. Indeed, it was knowledge beyond the capacity of modern man! Kelly Segraves reminds us: "With all our intelligences, we cannot figure out how pyramids were constructed."[122]

We ask again, could this intelligence have been transmitted by the "sons of God"?

Egypt is not the only country that has pyramids. A whole network of them can be found around the planet—in Cambodia, Shen Shi,

China, Thailand, Mexico, Nazca, Yucatan, Alaska...How does one account for such advanced scientific technology, on such a universal scale, and in such prehistoric times?

Could the answer be man's longevity at that time? The average age span before the Flood was close to nine hundred years—thirteen times the proverbial "three score and ten." This, of course, provided exceptional opportunity for learning, research, experimentation and the accumulation of knowledge. If only modern-day scientists could live that long! But longevity alone could never account for the specialized knowledge that our forefathers possessed. Their awesome expertise indicates a source outside of themselves. No other viable explanation can be offered except that mankind was preprogrammed from the depths of space.

The Bible not only supplies the key to the source of this knowledge but also to the way it was transmitted—space-beings called "fallen angels." As we have seen, such extraterrestrials possess superior knowledge. This they transmitted to man in direct contradiction to God's wishes.

Satan, the leader of the fallen angels, is himself a creature of rare brilliance and inimitable genius. The Latin translation of his name, "Lucifer" (from the Hebrew *Helel*), comes from a root meaning "brilliance or "magnificence." This is a trait he shares with his cohorts. He and they have access to classified, divine information, and are cognizant of hidden things (See Acts 16:16, 17). According to Professor C. S. Dickerson, "The source of their knowledge is found in their superior created nature and in their vast experience, as they lived through many thousands of years observing and collecting information."[123]

John L. Nevius, after surveying demon possession in China, documents the amazing knowledge revealed by the demon-possessed. Many persons while demon possessed give evidence of knowledge which cannot be accounted for in ordinary ways. They often appear to know of the Lord Jesus as a divine person and show an aversion to, and a fear of Him. They sometimes converse in foreign languages of which in their normal states they are entirely ignorant. [124]

Betty and Barney Hill

Betty and Barney Hill of New Hampshire, while returning home from a vacation in Canada in 1961, spotted a flying saucer and pulled off the road to watch it land. The engine of their car went dead, but they have no recollection of what happened after that. The next thing they remembered was being close to home...Sixty miles south! After this bizarre experience, they suffered from nightmares, anxiety syndromes, and ulcers. They finally consulted a psychiatrist, the noted Benjamin Simon of Boston. By the use of hypnosis, he was able to induce from them—quite independently of each other—the story of those missing hours. Both told the same story. Taken aboard the flying saucer, they underwent physical examination by the humanoid occupants. Tape recordings were made of their story (given under hypnosis), and a book was published. Later, an NBC television movie was made of the event.

Stanton T. Friedman. A nuclear physicist, studied these reports and was greatly impressed. But what finally convinced him was the map drawn in 1964 by Betty Hill of a star system unknown to science at the time. Betty claimed that she had been shown this map aboard the UFO, and revealed the details of it under hypnosis. The astronomers who examined the map at that time, dismissed it. But since 1964 new evidence has appeared on the subject—star systems that were invisible in 1964 have now been discovered. And an amazing fact has come to light. Dr. Friedman explains: "Using these new data, a computer came up with a map of the Zeta Reticuli system—faint stars 220 trillion miles away—which astronomers agree matches Mrs. Hill's precisely."[125]

Giants in Wickedness

Giants in wickedness is another distinction of the Nephilim. Sired by demonic beings, their character and activity certainly reveal the nature of the "fallen ones."

Many of the legends surrounding the cross breeding between the natural and the supernatural depict subhuman behavior. To the last one, the semi-gods of mythology—Zeus (the Roman Jupiter), Poseidon (Neptune), Eros (Cupid), Hermes (Mercury)—were violent, wanton, lustful and promiscuous beyond restraint. They constantly engaged in sex orgies and seduction, and produced the strangest of offspring. Emil Gaverluk tells of Zeus:

> He disregarded marriage laws and engaged in love affairs with members of both sexes. Zeus married Hera, his sister. One of the loves of Zeus was Europa. He seduced her by becoming a bull and carrying her away. Another was Leda, daughter of Thestios, King of Aetolia and the wife of Tyndereus, King of Sparta. But this did not stop Zeus…Athene was the daughter of Metis by Zeus. Metis tried to evade him. He seduced her. She warned him that if he did this again and the child was a male, that child would depose and conquer him. Zeus didn't like the sound of this and took no chances. He swallowed the child whole.
>
> Zeus' amorous victories illustrate the actions of uncontrolled spirit-beings lusting after human flesh.[126]

This is only one example from Greek mythology of the evil associations between spirit-entities from space and human beings from Earth. But it was not only the spirit-beings who acted this way; they seem to have passed on the trait to their descendants.

The Nephilim, in this respect more than in anything else, were close imitators of their fathers. They reflected the works of their demon ancestry. And just as there are degrees of goodness among the saints, so there are degrees of evil among the demons. Evidence of this is recorded in Matthew 12:43–45. It tells of the unclean spirit returning to occupy the house from which he had been displaced, and returns with "seven other spirits *more wicked than himself*" (emphasis added).

Irrespective of the degree of evil, all demons are regarded by God as vile and depraved. A recurring biblical adjective is "unclean" (Matthew 10:1, Mark 1:27, 3:11). These spirits are both morally and spiritually unclean, and the same distinctive applies to their progeny on Earth. What is more, those who traffic with such beings, frequently end up like them. Indeed such is the degree of their degradation that the Lord abandons them to their own depravity. Romans 1:24–32 graphically states that "God gave them up." The depths of their degradation and the infamy of their immoralities, puts them on a level lower than animals. Abandoned by God to the consequences of their wickedness, and deranged by demonic harassment, people have been driven to insane asylums and even to suicide.

Is it any wonder when incorrigible spirit-beings and rebellious humans cooperated to pollute Earth, to seek genetic control and produce hybrids that threatened the race itself, and even tried to thwart the very plan of God, that God should intervene in a judgment terrible to behold?[127]

Giants in Pride

The Nephilim were giants in pride also. Lord Acton claimed that power corrupts, and absolute power corrupts absolutely. So does knowledge. With their superior intelligence and knowledge, the Nephilim soon succumbed to the sin that became the downfall of Lucifer—his followers also dreamed of being gods, desiring to control and rule the Earth. Their advanced knowledge was the desired diet for their egoism. Such knowledge wedded to an already arrogant nature led to self-deification. They craved divine honors and religious worship. This was the ambition of Lucifer, of the Nephilim, and of every fallen creature. They were propelled in all that they did by self-will, self-determination, self-glorification, and ultimately self-deification. No wonder Josephus refers to the Nephilim as "sons who were overbearing and disdainful of every virtue."

Man's fascination with the occult and other-world phenomena is amazingly significant today. At no other time has he been more conditioned to accept the possibility of extraterrestrial life. Having once escaped his Mother-Earth, man now strains at his celestial tether, wondering what—or who—lies beyond the final frontier. With this starward look, the stage could be set for the coming of the unwelcome guests from space. The "days of Noah" are again here; perhaps even now agents from another realm are plotting the return of the Nephilim.

9

Humanoids Make Their Entry

The second coming of Christ is vividly foreshadowed by the "second coming" of diabolism.

—D. G. Kehl

The burgeoning of demonic activity in our time is proof of our proximity to the "last days" and to the mightiest power encounter of all history. The "sons of God" are on a collision course with the saints of God, and we may well be witnessing the opening scenario.

"No matter where you live on this planet someone within two hundred miles of your home has had a direct confrontation with a frightening apparition or inexplicable 'monster.'" So writes John Keel, one of the world's leading authorities on psychic phenomena and unidentified flying objects. "There is a chance—a very good one—that sometime in the next few years you will actually come face to face with a giant haircovered humanoid or a little man with bulging eyes, surrounded by a ghostly greenish glow."[128]

Have They Arrived?

Does this mean that the "sons of God" have finally arrived? How much hard evidence is there that the Nephilim are now here? And are the Nephilim to be identified with the occupants of these space vehicles that circle our globe and sometimes land on our planet?

No question is more vital, and certainly none more intriguing, than that of trying to identify the occupants that emerge from these space vehicles. These entities have been seen on numerous occasions, sometimes inside and sometimes outside the vehicle. Added to the hundreds of known witnesses are a multitude of others who have decided not to reveal their observations for fear of ridicule.

A few years ago, UFOs themselves were scoffed at, but accumulating data and the indisputable evidence of so many witnesses has changed that. Today, Ufology is treated with a degree of acceptability and respect. As for UFO occupants, however, this question is still very much a "no no." Creatures from outer space are unknown to science. Consequently to study them would be to invite certain ridicule.

However, in view of the evidence submitted by hundreds of bona-fide witnesses, plus footprints and other physical records left on *terra firma,* the question of the humanoids can no longer be ignored. One is forced to ask: Who are they? From where do they come? Whither do they go? What is their mission? Why are they here? What are they like?

Before we attempt to answer these questions, we must first determine the number of times that humanoids have actually been seen by people. Are the aliens many or are they few?

Facts and Figures

In 1954 in *France,* forty-two cases of humanoids and sightings were recorded. In Latin America in 1965 there were twenty-five such recorded cases, many of them multiple-witnessed. This gives us a total of sixty-seven sightings in just two years and in France and *Latin America* alone. Add to these the other instances that have been recorded in the *U.S.A.,* the *USSR.,* the *UK,* and *Australia,* and the number comes to many hundreds—far too many to be unceremoniously dismissed as mere folklore.

Another factor that should not be ignored is that these witnesses, although living in different parts of the world, where most had never heard or read about the others, describe the humanoids with an uncanny similarity. And much of this similarity is of such a character that only an expert in the field could understand and appreciate it. Collusion between the witnesses was obviously out of the question—a further factor which confirms their credibility.

Dr. J. Allen Hynek refers to this when he says:

> The reader will discover for himself that there is a very great similarity in accounts of occupant cases in reports from all over the world. He will learn that they are similar not only in the description of the appearance of most humanoids but in their reported actions...
>
> It would be helpful, one feels, if we could demonstrate that Close Encounters of the Third kind differ systematically from the other five UFO categories, Then we could, with some comfort, dismiss them. But they do not differ in any way—by geographical distribution, by times of occurrence, in numbers, and especially in **kinds** of observers—except that the relative number of cases with multiple witnesses is somewhat less...[129]

We have already discarded the thesis that the humanoids are denizens of other planets or galaxies, and thus limited our options to one of two. These entities are either spirit-beings (angels) from Heaven or spirit-beings (demons) from Hell.

As to which of these two options, I shall reserve judgement until the next chapter. At the moment let us examine the available physical descriptions of these humanoids to find a clue to their identity.

Physical Features

The variety of their physical features us such that Donald B. Hanlon may be right when he says, "It would seem at first glance, as if our 'visitors' had made a deliberate and concentrated effort to confuse us."[130] As to *size,* they range all the way from dwarfs to giants. Some were smaller than a meter in height. The humanoid that Mme. Leboeuf saw in Chabeuil in 1954 was like "a child in a plastic bag." The humanoids seen by Professor R. L. Johannis in the famous "Villa Santina" case in 1947, were no more than ninety centimeters high. The ones seen by Yago and Payo Rodriguez near the National Polytechnic Institute in Mexico City in 1965 were only about eighty centimeters. And the two humanoids seen by the gold-miners John Q. Black and John Van Allen near Bush Creek, California in 1953 were simply described as "midgets." Of the twenty-five sightings in Latin America in 1965, ten of the humanoids were described as "small men" and twelve as "tiny."

On the other hand, some of the creatures have been described as veritable giants, measuring anything up to fifteen feet! Those seen in 1962 by Dr. Gazua, a well-known and highly reputable physician in Crespo, Argentina, were described as "gigantic beings." The testimony presented by this famous doctor electrified the country. The humanoids observed in 1952 by a number of people in Flatwoods, West Virginia, were described as between ten and fifteen feet tall. The two humanoid footprints seen in 1963 by John McGoldrick and his friend in Sandling Woods, England, measured two feet long and nine inches across. Between these two extremes there have been many humanoids described as similar to human beings in size.[131]

More Features

As to the *other features,* they present the same kind of variety. Most have been described as having two eyes—usually very large ones—but there have been a few described as one-eyed, and some as three-eyed.

As to the *complexion*, some were seen as white and pallid, others as blood red. Some were hairy, others had no hair at all. One humanoid was described as having "wings like a bat," another as a "moth-man." Some were seen as "Christ-like beings" and others as monstrosities of evil.

The most consistent of humanoid descriptions speak of the similarities between humanoids and humans. Some as subhuman, others as superhuman, but in both cases much closer to human-beings than to any other living creature. Erick Von Däniken says the same thing but in a different way: Not that they resemble us, but we them!

Just as in our time humanoids resemble humans, so in the Bible times angels resembled humans. Captain J. C. Brown, late editor of "Living Links" makes this interesting comment: "Whenever in the Bible we read of angels contacting people, three things were always noted about them. First of all, they suddenly appeared; there was no approach from a distance. Secondly, **they always were taken at first for men.** (See Genesis 18 and 19; Judges 13:2,3; Kuje 24:4; Acts 1:10). Thirdly, they always brought some rational comforting message from God."[132]

As for details of the humanoids, a number of witnesses have referred to their large heads and craniums. In the famous contact made by Barney and Betty Hill of Boston, the humanoids had odd-shaped heads with a large cranium which got smaller in size toward the chin. In the Villa Santina case in Italy, Professor Johannis described the two humanoids in these terms: "Their heads, according to the impression I got, were bigger than the head of a normal man. (Although in size they were no taller than 90 centimeters)."[133]

Aimé Michel describes the little men seen in the Provencal Village of Valensole, France, as "pumpkin-heads" and "with brains at least three times as big as ours." In sighting after sighting, in South America, Mexico, U.S.A., France…there seems to be a preponderance of small humanoids with big heads.

Another feature that has received increasing attention is their eyes. Witness after witness has commented on the size, type and position of the eyes. Here are a few specimens drawn from dozens of such cases.

Barney and Betty Hill recalled under hypnosis that the humanoid they saw had slanted eyes that continued around the sides of the head. In this way they seem to have wider lateral vision than humans.

Professor Johannis states that the eyes of the two humanoids he saw were enormous, protruding, and round. He compared their appearance and color to two well-ripened yellow-green plums! And he noticed something else: "In the center of the eyes I noticed a kind of vertical 'pupil.' I saw no traces of eyebrows or eyelashes, and what I would have called the eyelids consisted of a ring, midway between green and yellow, which surrounded the base of those hemispherical eyes just like the frame of spectacles."[134]

Antonio Villas Boas, describing the female humanoid aboard the vehicle where he was forcibly taken, tells us that her eyes "were large and blue, more elongated than round, being slanted outwards (like the slit eyes of those girls who make themselves up fancifully to look like Arabian princesses; that is how they were, with the difference that here the thing was natural for there was no makeup whatever.)"[135] Charles Hickson and Calvin Parker of Pascagoula, Mississippi, described the three humanoids they saw as being silver-skinned and "with big eyes."

It is surprising how many witnesses have mentioned how thin *the lips* of the humanoids seemed to be. Barney Hill said they were "much like when you draw one horizontal line, with a short perpendicular line on each end. This horizontal line would represent the lips without the muscle that we have."[136]

Antonia Villas Boas also commented on this fact and reported that the female he saw had lips that were thin and hardly visible.

The face of the humanoid narrows considerably and ends in a

pointed chin. This was mentioned by Antonio Villas Boas, Mrs. Wesley Symmonds of Cincinnati (Ohio) and many others.

What adds to the interest is that all these features—heads, eyes, lips, and chins—are combined in a number of important cases. M. Maurice Masse of the French Provencal village of Valensole, describes the entities he saw on the morning of July 1, 1965, as having "pumpkin like heads, high fleshy cheeks, large eyes which slanted away around the sides of the face, slit mouths, or 'holes' without muscular lips and very pointed chins."[137]

Mrs. Wesley Symmonds of Cincinnati, describing her experience of July 3, 1955, as she was driving through Stockton, Georgia saw a humanoid that "had large, bulging eyes, a cap-like affair on his head, no visible mouth, a long pointed nose, and a chin that came to a sharp point…Its long, thin arms ended in claw-like appendages."[138]

And, of course, the granddaddy of all occupant sightings—the one witnessed by the Sutton family in Hopkinsville, Kentucky. Bill Taylor, a relative, was visiting the Sutton home the evening of August 22, 1955, and describes one of the entities he saw: "It appeared to be lit by an internal source, had a roundish head, huge elephantine ears, and a slit-like mouth which extended from ear to ear. The eyes were huge and wide-set. Only about 3 or 3 ? feet in height, the creature had no visible neck, and its arms were long and ended in clawed hands."[139]

In the last two cases, mention is made of *clawlike appendages* instead of hands. This also seems to be the description made by a number of other witnesses. Jacques Vallee mentions one of these when he writes of three young men in San Carlos, Venezuela, seeing a small humanoid on December 16, 1954. In spite of his dwarfish size, this humanoid attacked one of the young men with his *claws.* The young man was later treated in the hospital for his wounds.[140]

It is interesting how the color "red" appears in many of the descriptions. The group who saw the giant humanoid in the Flatwoods, West

Virginia, on September 12, 1952, mention his "blood red face." Antonio Villas Boas describes the hair on the armpits of the woman with whom he had relations aboard the space vehicle as "very red, almost the color of blood." Barney Hill described one of the humanoids he saw, and says that he reminded him of a "red-headed Irishman." The entity seen by the three boys in Belo Horizonte, Brazil, on the evening of August 28, 1963, had a complexion that was a "vivid red." The entities seen by Eduardo Lujan Yacobi and his wife on August 20, 1965, Argentina, were "shining violet-colored or reddish-colored figures."[141]

Like and Unlike

What do all these descriptions really tell us? One thing for certain: *These entities are like humans, and yet quite different.* According to the laws of biology the human body is typical of planet Earth. The Bible corroborates this in 1 Corinthians 15:38–40, 44–49. This "earthly" body is adapted to the gravity, pressure and chemical composition of the atmosphere. And it is conditioned by factors like our distance from the sun. Beyond that, it is not possible to be definite, and men can only speculate. Thus we are left with the fact that the physical nature of these beings tells us little of their true identity. They share some similarities with humans, but that is all. To identify them with more precision, one has to examine their character. Are they benevolent or malevolent? Friends or foes? Are they here to help us, or to corrupt and seduce us?

10

God's Angels or Luciferian Forces?

. . . the still vex'd Bermoothes.

—William Shakespeare

In spite of similarities, humanoids can never be explained or understood in solely human terms. Some researchers go a step further: humanoids cannot be explained in terms of other planets and galaxies, either. This leaves us with only two options: They are angelic beings, and thus benign; or demonic entities, and thus malign. But which one of the two?

Saviors from the skies, or demons from the depths? Do they resemble the good aliens of Spielberg's "Close Encounters" or the bad Martians of Wells' *The War of the Worlds*? God's angels or Luciferian forces?

If their physical characteristics do not supply the key, then we must look at their moral attributes to determine their identity.

Benign Beings?

Some assert that they are benevolent. These visitors are our friends. They are engaged in nothing less than a mission to help us. The Saucer People are our first line of defense, there to protect us. They are a "galactic peace corps" to save us from ourselves. To argue their case, advocates of this theory underscore the following items:

- There is no known case where these beings have deliberately caused the death of anyone. A powerful argument, if true. But, as we shall see later, an argument that is open to serious doubt.
- There is no known case where one of their spacecrafts has ever collided with a vessel from Earth. Again, open to doubt.
- There is no known case where space-beings have ever attempted to invade and subdue our planet, although undoubtedly possessing the power and "know-how" to do so.

To bolster these arguments, we are further reminded of the occasions when these beings have been friendly toward the humans they encountered. They have even "waved" to them and invited them aboard their craft! What is more, these beings at times have revealed a shyness and reticence that made them retreat into their craft when humans rejected their invitation! It has even been suggested that they are not only friendly in their intentions toward us, but that they are actually here to save us. They wish to warn us of impending calamities and catastrophes, especially if we continue playing with our nuclear bombs and missiles. What other reason could there be for the great increase of UFO sightings since 1947? It was just before then, that atomic bombs were dropped on Hiroshima and Nagasaki. The carnage and destruction was such that these alien-beings rushed here to warn us.

John Keel writes:

> On August 3, 1958, HAM radio operators throughout the U.S. reportedly picked up a strange broadcast on the 75-meter international band. A male voice purporting to be "Necoma from the planet Jupiter" warned his listeners that the American bomb tests could lead the world to disaster.

> He spoke for **two and a half hours** in English, German, Norwegian, and his own language, which was described as a kind of musical gibberish.[142]

Dr. Clifford Wilson elaborates: "The theory is that this (atomic explosion) forced blasts of energy out into space, detected by sensitive instruments of super-intelligent beings observing us from other worlds. Investigative expeditions have been sent to Earth to warn the foolish inhabitants about the danger of continuing such activities. According to those who argue in this way, messages given to contactees have centered on problems Earth is facing if it continues to experiment with forces of nature in this way."[143]

Space Brothers?

Some explain that the great balls of light that these extraterrestrial beings send out into the atmosphere are actually there to neutralize dangerous atomic radiation resulting from man's experiments. Since human scientists insist on poisoning Earth's atmosphere, these super-intelligent beings are engaged in minimizing their effect. Indeed, it has been suggested that their concern for us is such that they have already assembled large fleets of UFOs in order to evacuate people from this planet before some dreadful catastrophe occurs. "Many ufologists claim that spacemen are standing by with thousands of machines, ready to save enough of their Earth-colonists so that they will be able to re-launch an earth-type civilization in another space home already prepared for them."[144]

Some Bible students have picked up this idea, interpreting such an invasion as the Rapture! Furthermore, they interpret the messages received by some of the contactees as parallel to the predictions of Scripture. Indeed, their coming to Earth in itself has been interpreted as fulfillment of prophecy. And even those who have ceased to believe in divine intervention seem to have no difficulty in believing in extraterrestrial intervention!

Add these facts together, they tell us, and they demonstrate quite clearly that these beings from space are indeed our friends, and may one day be our deliverers.

Ted Owens of Cape Charles, Virginia, claims that he is Earth's principal intermediary for Space Intelligence matings, assisting mankind through the centuries. He is the man who got a lot of publicity in the May 3, 1979, issue of *Sports Illustrated.* Owens decided to cast a spell upon the Baltimore Colts, and then charge $100,000 to unhex them! "Now" asked *Sports Illustrated,* "was it Owen's fault that Johnny Unitas tore his Achilles' tendon, Tom Matte got appendicitis and Sam Havrilak sprained his ankle?" Later, when a Philadelphia writer challenged Owens to prove he could do something, the PK man announced he would hex Tom Woodeshick of the Eagles, and within fifteen minutes, Woodeshick was dismissed from the game for fighting.[145]

If one believes in this sort of thing, obviously these powers are not uniformly benign! But far more important than these speculations, are a number of formidable facts that have to be taken into account—facts that could well convince us that extraterrestrials are not benign, but hostile.

Hostile Visitors

Let us take a closer look. Consider first the deliberate cases of outright abductions that have happened from time to time. Mark Albrecht and Brooks Alexander in their excellent article in the *SCP Journal* (August 1977) refer to what can be considered a prototype of abduction encounter. It happened in 1971 near the Superstition Mountains of Arizona to a man called Brian Scott. He was on a camping trip when he found himself levitated into a large UFO. In the report of the doctor who directed Scott's hypnotic regression, he states that Scott had an authentic and detailed memory of what happened.

The subject finds himself...In a small room, paralyzed and unable to resist. Suddenly several 7-foot-tall beings enter and undress (him)...The beings are ugly, with sloping shoulders, crocodile-scaled skin, elephant-like feet, and hands with three fingers and a recessed thumb...Heavy fog or mist is everywhere...Two of the 7-foot-beings station themselves at consoles of some kind, and a third stands beside a pole on which there is a moveable box with many tiny colored blinking lights...The subject experiences a series of uncomfortable if painless sensations from his feet upward; he senses he is bleeding; he urinates; he feels water run from his stomach; he feels his chest opened and he thinks his heart has left his body briefly; finally head feels "pulled" violently...The subject detects a distinct unpleasant odor. Then from across the room out of the fog comes a 9-footer... (and) apparently communicates telepathically with the subject without moving his mouth. A message is communicated to the subject. Then the subject experiences an out-of-body trip. The message is a combination of vague philosophical statements, general information about the alien's origin and purpose, and a promise that they will return.

Space-Nappings

Worse than abductions have been the cases of "space-nappings." There is evidence of a number of adults, and even children, who have disappeared, never to be seen again. Here is just one example:

Lieutenant Cody and Ensign Adams were two Naval officers who disappeared off the Californian coast in 1942. They were in a small blimp at the time, watching out for enemy submarines. The crews of a number of patrol boats and fishing smacks in the area saw something happen to the blimp that was incredible: they saw it suddenly shoot straight up into the air in a manner that was physically impossible. At about 2,500 feet, it leveled off, and then drifted aimlessly toward the coast where it finally touched ground about two and a half hours later. Everything inside the gondola was in perfect order—except that the two officers were not in it. And they have never been found since. What is strange is that the blimp had been under close observation all this time, and the men had neither fallen nor jumped out. Nonetheless, they disappeared.

From an entirely different source comes this intriguing admission by the foremost Russian authority on UFOs, Dr. Felix Zigel. He contends that UFOs may have "frightened, harassed and possibly even killed Russian cosmonauts on their mission"[146] It is obvious that they are not respecters of persons or nations!

Another frightening feature resulting from contact with extraterrestrials, is the number of contactees who have died shortly afterwards—some of them from leukemia, a disease that can be caused by radiation poisoning. What compounds the tragedy is that a number of these contactees were comparatively young. Hardly an act of friendliness!

Major Donald E. Keyhoe lists a whole catalogue of incidents where UFOs have been responsible for deaths and injuries to humans. And John Weldon supplies further documentation accusing UFO occupants of being "responsible for aggravated assault, burnings by direct ray focus, radiation sickness, murders, abductions, pursuits of cars, assaults on mobile homes, paralysis, cremations, disrupting power sources, etc."[147] And to leave us in no doubt whatsoever, he adds that even UFO investigators have succumbed to sudden and mysterious deaths, others to suicides and nervous breakdowns. Some would add to this list the horrible mutilations of animals witnessed in a number of midwestern states. In the eighteen month period leading up to January 1977, there were seven hundred such mutilations. Most of these occurred at nighttime and in isolated areas. No cause of death could be determined; no trace or tracks of the killer could be found;

and in many cases the animals seemed to have been dropped from the air. The sight of the mutilations were such that they froze the blood of those who found them. Hardly the work of benign space brothers.

Mid-Air Collisions

In spite of the claim made that there has never been a collision between a spacecraft and an Earth vessel, Robert D. Barry in his lecture "From Outer Space to Earth" cites two such accidents. Both cases involved C-118 transports. These collided in mid-air with UFOs, but little publicity was given to the fact.

The first collision occurred on April 1, 1959, near McChord Air Force Base in Washington. In this incident, the pilot, and the two crew members were all killed.

The second collision was in 1960, over Wisconsin. There were seven persons aboard the plane, but only two bodies were ever found. The other five had disappeared into thin air.

It is significant that in both collisions, there was apparently no damage to the UFOs.

To these two collisions, one could add the case of the F-89 jet interceptor that was dispatched from Kinross AFB (Michigan) on the night of November 23, 1953, to intercept a reported UFO in the area. Lieutenant Felix Moncl Jr. was the pilot and Lieutenant R. R. Wilson his radar observer. They were guided to the UFO by Ground Control Intercept radar station. They reported back to Ground Control that they had spotted the UFO, and described it as circular in shape, and bright bluish-white in color. They then tailed the UFO over Lake Superior at 500 mph. Minutes later, the Ground Control Intercept controller was shocked to see the blips of the AF jet and the UFO suddenly merge. The F-89 and the UFO were locked together. Moments later both vanished from the radar screen. And there has been no trace of the two airmen or of the F-89 ever since.

More recent was the case of the Cessna 182, cruising over the Bass

Strait, some 130 miles south of Melbourne, Australia. This was on the night of October 21, 1978. The pilot, Frederick Valentich, suddenly saw an unusual UFO approaching from behind, and a few minutes later he and his aircraft had disappeared for good.

When Valentich first saw the mystery spaceship, he radioed the Melbourne Flight Service Control and described his pursuer as "a green light and sort of metallic light on the outside." As the object came closer, he radioed that he could see "four bright lights." Some of his last words were: "It isn't an aircraft…Melbourne…it's approaching from due east toward me…flying at a speed I cannot estimate. It is a long shape. Cannot identify more than that…Coming for me right now…the thing is orbiting on top of me." He then reported that the engine of his plane was "choking." Metallic scratching replaced the pilot's voice. Complete silence.[148]

All these four cases involved direct mid-air collisions between airplanes and UFOs. In addition, there are cases which report the complete disappearance of earth vessels, leaving no trace of them. Both airplanes and ships have disappeared never to be seen or heard from again.

The Bermuda Triangle

Most people have heard from time to time of the mysterious happenings within the so-called Bermuda triangle. Number of planes and ships have disappeared in this area which is bordered by Bermuda, Puerto Rico and Florida. So significant is the number of these missing vessels that the triangle has earned for itself a number of eerie designations: "The Devil's Triangle," "The Limbo of the Lost," "The Hoodoo Sea," "The Twilight Zone," and "The Port of Missing Ships." Many different explanations have been advanced as to the cause of their disappearance, but none is more persistent than that of hostile UFO activity. As a matter of fact, many ufologists consider the Bermuda Triangle as a "show piece" area for major UFO activity. It has been claimed that large space vehicles have appeared and "swallowed up" these vessels together with the earthlings on board.

The most publicized disappearance of all was that of a squadron of five Avenger torpedo bombers on the afternoon of December 5, 1945. With a complement of fourteen crewmen, this squadron flew out of Fort Lauderdale Naval Air Station in Florida, on a routine Atlantic flight. The war was over by this time, so there were no enemy planes anywhere in the vicinity. Weather conditions were reported as excellent. All five planes were in good mechanical order, including all instruments, equipment, and engines. Each plane had been carefully preflighted and each held a full load of fuel. Each plane also carried extensive radio gear, including ten communication channels, plus homing devices. In addition, each plane had a self-inflating life raft, and each man wore a Mae West life jacket. Finally, all pilots and crewmen were experienced. And to top it all, it was a comparatively short flight—about two hours in all.

But something mysterious happened to all five planes. When they should have been arriving back at the air base, there came instead a strange radio call from the flight leader. He reported that the sky above and all around them appeared strange and that they had no sense of direction whatsoever. They were directed to "head due west." After a long silence the patrol leader came on again, with alarm in his voice: "We don't know which way is west. Everything is wrong...strange...we can't be sure of any direction. Even the ocean doesn't look as it should!"

About forty minutes later there followed another garbled report in a hysterical voice stating that weird aircraft were closing in on them. After that it was complete silence. All five planes had disappeared leaving no trace.

A giant Martin Mariner flying boat, with a crew of thirteen, was sent out to search for them. The plane was equipped with full electronic search devices and had the capability to land on the roughest of seas. But this plane has also disappeared into the grim silence of the Atlantic.

After the disappearance of the sixth aircraft and their twenty-seven crew members, there followed the greatest search operation in history. For five days an armada of three hundred planes and twenty-one ships crisscrossed the sea and the sky, but the grim Bermuda Triangle yielded none of its secrets. To this day it remains a mystery as to what happened.

One wonders with Art Ford, the radio newsman from Florida when he said, "The planes vanished in an area, which is now our space-shot center. The same reasons that make the East Florida Coast ideal for rocket shots into space might also make it ideal for landings from outer space."[149]

All one can be sure of is this: That if the UFOs are responsible for these disappearances, there goes the theory of their friendliness!

Major Donald E. Keyhoe in his book, *Flying Saucers from Outer Space,* asserts that UFOs show a great interest in military establishments and in defense areas. There have been numerous sightings over atomic energy plants, Air Force bases, Naval bases, Marine Air Corp. stations, aircraft plants, rocket-testing bases and uranium mines. Dr. Clifford Wilson concludes that, "There is a frightening possibility that UFOs are especially active in specific centers such as these because of their military potential for future operations."[150] Put that down as another "unfriendly" act from our visitors.

The White House, D.C.

In this context one should include the sightings of UFOs over the White House in Washington, D.C. This happened on two specific occasions: The first on the evening of July 6, 1952, and the other, six days later. Officials in the towers at St. Andrews Field and the Washington Airport confirmed the sightings, as did a number of airline pilots. Major Keyhoe underscores the potential seriousness of this particular UFO activity: "Up there in the night some kind of supermachines were reconnoitering carefully. From their controlled maneuvers it was plain that they were guided— if not manned—by highly intelligent beings. They might be about to land—the Capitol would be a logical point for contact. Or they might be about to attack."[151]

One more question needs to be asked regarding the entities who man the UFOs. Is there something about their attitude that could supply the key to their character?

Although a few of the contactees have said that the humanoids they saw appeared to be friendly, quite a number have indicated otherwise. To them, the humanoids appeared unfriendly, even hostile. Witnesses in the Kelly-Hopkinsville case were harassed and frightened over a period of many hours, revealing persistency on the part of the humanoids in prolonging human suffering. More than one witness has referred to their attitude and behavior as "animalistic."

Indicative of their presence in a large number of cases is the strong and repulsive stench they leave behind. Frequently this irritating stench lingers on into the next day, as in the case of the Flatwoods incident.

They Carry Weapons

Even more indicative of their hostility is the weapons they carry. It would seem at this stage that their aim is not in killing their victims but rather in immobilizing them by paralysis. Witness after witness has referred to flashes or rays from these humanoids (sometimes from their belts, sometimes from their chests) that rendered them paralyzed—at least temporarily. Professor Johannis in the Villa Santina case writes: "What is certain is that one of them raised his right hand to his belt, and from the center of the belt there came something that seemed as though it might be a thin pull of smoke. I now think it was a ray or something of the sort…Anyway, before I had time to move or do anything, I found myself laid out full length on the ground. My pick shot out of my hand, as though snatched by an invisible force."[152]

In the Cisco Grove case, the witness was gassed by a special kind of smoke emitted by the humanoid: and each time this happened, the witness passed out. When he awoke later, he found himself sick and retching.

In the Valensole case in France, M. Masse testified that when the humanoids he saw pointed a pencil-like instrument at him, he was immediately stopped in his tracks, unable to move. In a similar fashion, Jose Parra of Venezuela found himself totally immobilized by a "violet-colored beam" from a small device which one of the six humanoids pointed at him. One of three boys in Belo Horizonte, Brazil, tried to throw a brick at a humanoid when it had its back towards him. But the creature veered round "shot an orange beam at him from a square lamp on his chest, and 'paralyzed' the boy's arm."[153]

There are innumerable other instances from the United States, Latin America, Europe…where witnesses were paralyzed by a "ray" that rendered them completely immobile.

They Quote Scripture

Some have difficulty in accepting the fact that these space beings are malevolent because some of the predictions they make parallel biblical prophecies. They have been known to quote from the Bible and to have shown remarkable interest in scriptural references to the last days. Armageddon is number one on their priority list.

It is even claimed that some of the content of the Bible was brought to this planet by extraterrestrial beings. Dr. Irwin Ginsburgh, a leading physicist, suggests that the biblical description of hell could well be a description of the Planet Venus, and a description transplanted to Planet Earth by space beings. It has been discovered that Venus has a searing atmosphere, with a surface heat of 850 degrees Fahrenheit. The planet is wreathed in dense sulphur clouds, with sulphuric acid droplets—liquid and solid, dripping from the clouds. Could our planetary neighbor be the origin of our hell? Dr. Ginsburgh thinks it could be: "It could be that knowledge of the surface conditions on Venus were communicated to primitive Earth people by extraterrestrials who showed them pictures of the planet's surface...and the dark, glowing, searing, hot and sulphurous planet has been remembered—as hell."[154]

If some of their predictions do parallel those of the Bible, how can one claim that they are malevolent? Would not the opposite be the case?

Surface Resemblance Only

The answer to this question is three-fold: First, although periodically there have been religious overtones to their messages and even a surface resemblance to Christian revelation, the over-all picture is certainly not one of friendliness. "Even when it seemed that genuine 'revelation' had been given, the evidence indicates that a pattern of deception has followed those earlier minor 'prophecies' that have been of a fairly general nature. Even when those 'prophecies' have been more specific, it would seem that they have been given as an inducement to cause greater belief in these strange beings. They in fact appear to be bent only on the fulfillment of their own plans, with no real interest in the welfare of those whom they contact."[155] The mere fact that these beings quote Scripture does not prove that they are benevolent. Satan also quotes the Bible—often for a seemingly benevolent purpose! When he suggested that Jesus turn the stones into bread, what could have been more humanitarian! For a person who had not seen bread for forty days, there could be no greater virtue! Quoting Scripture and partially fulfilling Scripture is no authentication of character.

Second, God's Word makes it clear that the mark of a true prophet is not one who makes an occasional right prediction but rather one who invariably makes the right prediction. There is to be no exception. Deuteronomy 18:22 states categorically that an unfulfilled prediction does not have its source in the Lord. God's standard is 100 percent accuracy. One inaccurate prediction and the man must be classified as a false prophet. It is obvious that not a single one of our modern physics would qualify for the biblical title of prophet. Indeed their degree of accuracy over a period of any one year would qualify for an F grade!

Here is an example of a famous prediction made by an UFO contactee back in the sixties. It stated that New York City would be submerged under the ocean on July 2, 1967. This date was accepted as genuine by UFO contactees and also by members of the Hippie community of New York City. But New York is still standing—submerged by an economic tide perhaps but certainly not by the Atlantic Ocean. After considerable research, more than one author has concluded that a *large part* of the data supplied by space beings and their contactees is deliberately fabricated.

Denying the Gospel

Besides all this, of course, is the theological argument. If these spirits beings are good angels, on the side of God, why do so many of their messages contradict the Gospel of Jesus Christ? And why do they contradict spiritual doctrines that have been considered cardinal by the Christian church for the past two thousand years? Why do they advocate such unbiblical propositions as: belief in an impersonal God; endless improvement in the hereafter; men are not lost sinners in need of divine mercy; Christ was divine only in the sense that all men are alleged to be divine; the Cross was not an atonement for man's sin; Christ's resurrection was a mere materialization; man's hope lies in human works and not divine grace.[156]

These messages clearly indicate a non-Christian source.

"By Their Fruit..."

Besides negative divine truths and doctrines, many of the messages received have espoused behavior and practices that are contradictory to the Christian ethic. They have espoused the use of LSD, condoned pre-marital sex and advocated racism.

It is obvious that these messages "from space" can in no way be considered Christian; indeed, they can only be regarded as having proceeded from the "Father of Lies" himself. "However good and beneficial the spirit-messages may be thought to be, the Gospel of Christ does not have priority in them. Secondary matter, if not outright blasphemy and obscenity, have usurped the place that rightfully belongs to the Gospel."[157]

One could proceed from this statement and add that not only is the Gospel of Christ denied priority in the message of these space beings but that they have never revealed the least interest in propagating the Gospel, nor the least desire in seeing the salvation of lost sinners. In the Bible, angels were actively and emotionally involved in such work. As a matter of fact, we are told that they rejoice "over one sinner that repents" (Luke 15:10). Modern UFO occupants have never engaged in such function or emotion.

Mr. Ashtar

Third, they have at times inadvertently betrayed their own lineage. Many of the contactees (and mediums for that matter) claim to have received messages from a "Mr. Ashtar." Who is this personage who calls himself Mr. Ashtar? We seemed to have met the name before.

Except for the gender, Ashtar or Ashtaroth, (Ashtoreth), was a female Canaanite Goddess. She was known in Babylon by the name of Ishtar, and in Ethiopia by Astar. She was recognized as a goddess of fertility and life, and her worship was characterized by lewd and licentious behavior. It was one of those religions which perverted its adherents. The tree of life became a phallic image. Such were the obscene orgies and vile practices associated with the cult of Ashtar, that it became *an abomination unto the Lord.* Nowhere in the Old Testament does God's anger reach such a boiling-point as when He decreed the complete extermination of the Canaanites. This race of people, many of them descendants of the infamous Nephilim, brought the land of Canaan to an all-time low in moral degradation. Such was its condition that God could no longer permit the situation to continue; their extermination had become a moral necessity. The bottom line in all this is: Ashtar was the goddess of these people!

Other Distinctives

In trying to assess the true character of these spirit beings, there are other factors that must be taken into account:

Their preference for darkness. By and large, the humanoids that have appeared to Earth people have been nocturnal visitors. They have avoided the light on every possible occasion. Jaques Vallee tells us that "only a vanishingly small number of landings take place during the day…"[158] The peak time seems to be around 10:30 at night. Others have been seen during the night; but hardly any sightings have been reported after 6:30 in the morning. This period of time, between 10:30 p.m. and 6:30 a.m. is when most inhabitants of Earth, of course, are fast asleep! This fact leads to another distinctive of our night-visitors:

Their preference for lonely and isolated places. With few exceptions, they seem to shun populous centers and contact with earthlings. Jacques Vallee tells us again regarding the numerous landings in France in 1954 that "the geographic repartition of the landing sites in 1954 is inversely correlated with population density."[159] It is interesting that Rabbi Hanina in the Mishnah tells us that demons "do not lurk in pits near centers of population."[160]

This combined aversion to light and publicity reminds one strongly of the Gadarene demoniac. John Livingston Nevius has an intriguing sentence germane to this point in his book, *Demon Possession.* Reviewing cases of demon possession in China, he says that they are very rare in large cities, and that they occur principally in rural and mountainous regions.

The same distinctive is found in the scriptures. We have no record of a case of demon-possession in the city of Jerusalem, and only one in Capernaum. Most of them were in the more rural settings of Gadara and Galilee.

Their reluctance to reveal themselves, even when seen by Earth people. Their tendency on being spotted is to disappear immediately. Charles Fort asserts that "the greatest mystery of all is this: Why don't they show themselves to us openly?"[161] This factor has certainly intrigued and mystified UFO researchers. Why are they afraid of revealing themselves? Why don't they want to be identified? Why do they shun the light? If indeed they do come from another planet and were remotely similar to us as humans, then why don't they want to reveal themselves? Instead they prefer the cloak of anonymity, isolation and darkness.

The weapons they carry with them may be indicative of the same trait. Usually they are seen to carry tubes, wands, chains, sticks, flashlights...According to Donald B. Hanlon these weapons could have been "designed specifically as a deterrent to any type of spontaneous intelligent communication between the witnesses and the occupants."[162]

An unpleasant odor seems to be their "calling card." Witness after witness has referred to the stench, irritating and nauseating, that frequently accompanies the presence of humanoids. And this stench lingers long after the humanoid has departed. One of the earliest references to a spacecraft is probably that found in an ancient Egyptian account, some 1,500 years before Christ, and one of the characteristics of that craft was the foul odor that issued from it.

In one of the bestsellers of the seventies, *The Amittyville Horror,* based on a factual account of what happened to a family in Amittyville, New York, the same characteristic of a nauseating odor seemed to accompany the presence of the "ghost" or "spirit-entity" that appeared from time to time.

The tendency of these spirit beings or humanoids is to obey the same law or code of instruction. Although many in number and different in certain features, yet they all seem to be under one and the same command.

For those who subscribe to the view that these entities represent the denizens of other planets, they would have to confine them to one planet only, since they all seem to obey a uniform command. If they come from a number of galaxies, and thus represent different cultures and lifestyles, one would not expect to find such uniformity of behavior. But the fact is that all these humanoids seem to act under the same set of rules. Aimé Michel has expressed it thus: "If the "X" system is a multiple one (if there are several origins or responsible parties), then they all obey equally, in so far as our observations permit us to gauge, **one single law on one precise point** and that is abstention from contact."[163] And, of course, there may be many other such laws on which they act in unity. All this leads us to believe that all the spirit entities that have visited our planet keep the same laws and obey the same leader. And this is precisely how Satan operates. All his cohorts obey him implicitly. Whenever or wherever they appear, they are under his over-all authority. Satan is the most unquestioned authority in Hell's chain of command.

There is an amazing similarity between the descriptions given of these humanoids and the traditional descriptions given of demons. Perhaps the best known drawing of a humanoid is that by Pauline Bowen, from a sketch released by the United States Air Force, and based on details supplied by the Kelly household in Hopkinsville, Kentucky. Studying that drawing, one can only conclude that it is more than coincidence how closely this drawing resembles certain medieval drawings of the devil. There is the forbidding black color and all the grotesque features. And one could add the reeking sulphur as well!

The clinching argument may well be the fact that *many human-beings who have been exposed to UFOs and their occupant humanoids have become themselves demon possessed in the process.*

11

With What Body Do They Come?

For spirits when they please,
Can either sex assume, or both; so soft
And uncompounded is their essence pure.

—John Milton

For decades illicit sex has been Hollywood's best-selling merchandise. She has churned out an avalanche of movies saturated with every obscenity in the book. In their attempt to sell the bizarre, the savants of the celluloid empire have extended themselves to the ultimate in the macabre—the sexual union between demons and humans.

The significance of this ghastly portrayal of demonic sex is startling, for it links the future with our past. The strange marriage between the "sons of God" and the "daughters of men" in Genesis 6 is likely to be repeated in our time, for the New Testament predicts that some of the distictives of the days of Noah will be repeated at the end-time. (Luke 17:26)

The fact that the "sons of God" were able to mate with the "daughters of men" presupposes the "sons of God" had material body. If so, what kind of body did they have? When the Nephilim appeared in Genesis, they were the result of the mating together of the "sons of

God" and the "daughters of men." This presupposes that the "sons of God" had material body. If so, what kind of body did they have?

Traditionally, angels are considered to be disembodied intelligences, purely spiritual beings and devoid of human passions. Angels have no concupiscable appetites, asserted St. Thomas Acquinas. Many modern researchers agree with this traditional viewpoint, and believe that angels are pure intelligence, possessing no material form. One of these theorists, Dorothy L. Sayers, summarizes this view for us:

> The answer usually adjudged correct is, I believe, that angels are pure intelligence: not material, but limited, so that they may have location in space but not extension. An analogy may be drawn from human thought, which similarly nonmaterial and similarly limited. Thus, if your thought is concentrated upon one thing—say, the point of a needle—it is located there in the sense that it is not elsewhere: but although it is "there," it occupies no space there, and there is nothing to prevent an infinite number of different people's thoughts being concentrated on the same needlepoint at the same time. The proper subject of the argument is thus seen to be the distinction between location and extension in space.[164]

This theory ignores the fact that angels do sometimes appear in visible form and do sometimes have an extension in space. On many occasions they have assumed human form and manifested human traits. "Every feature of personality is predicted of angels," according to Lewis Sperry Chafer.[165]

Since the early days of the Christian Church, many questions have arisen as to the composition of angels and numerous attempts have been made to define and describe them. Some of the early Church Councils went so far as to claim that these angelic bodies were ethereal and fire-like. Others held views similar to ancient Jewish rabbis who believed Daniel 10:6 gave a full description of such a body. "His body also was like the beryl, and his face as the appearance of lightning, and his eyes as lamps of fire, and his arms and his feet like in color to polished brass..."

The Book of Enoch (XV, 4–11) tells us that good angels are beings of fire; and they cannot eat, drink or multiply; but the bad angels—for some reason—do have the capacity to do all these things: eat, drink, and multiply.

Robert Genevicz in his recent book, Guardians in the Dark, draws a similar conclusion: "One thing seems certain, the fallen angels have always been very interested in the reproduction process as it relates to man."

Angels Incarnate

The rise of Gnosticism gave impetus to the belief that angels could put on a material body. Centered on Egypt, Gnosticism emphasized the dualism of spirit and matter. Spirit was good: matter was evil, and the two were in constant conflict. Thus when angels took on material and bodily form, it was nothing short of a cosmic catastrophe. The angels of Genesis 6 were guilty of sin not only because of their lustful associations with the women of Earth, but for the very fact that they had donned a material form.

Platonists like Apuleius not only accepted the teaching that demons have material bodies, but proceeded to define these beings. They taught that demons were "of an animal nature, passive in soul, rational in mind, aerial in body, eternal in time."[166] This definition solicited the withering addendum of Augustine: "Those aerial animals (who) are only rational that they may be capable of misery, passive that they may be actually miserable, and eternal that it may be impossible for them to end their misery."[167]

Later in church history, the Scholastics and Lateran Council (AD 1215) decided that angelic or spirit-bodies were incorporeal and immaterial in nature.

Kenneth S. Wuest of the twentieth century, however, disagrees, and draws the same distinction between angels and demons, as the Book of Enoch. In Matthew 12:43, Jesus speaks of an unclean spirit who went out of a man, then walked through dry places, seeking rest.

But finding none, he determined to return to the house from whence he came out. Wuest comments on this: "This clearly infers that at one time they had physical bodies, and being deprived of them, through some judgement of God, they tried to satisfy their innate desire for a physical existence in that way. This is not true of angels."[168]

One does not of necessity have to subscribe to this theory, or course, in order to believe that angels (good or bad) can take on visible, physical form. Many examples can be found in Scripture where angels took on a physical, visible form temporarily. The unique feature about the fallen angels of Genesis 6 however is that they seem to have taken on such a form *permanently*—or at least, semipermanently.

Reticence of Scripture

Compared with the writings of many of the Church Fathers on this speculative subject, we cannot help but note the great reticence of Scripture. The existence and mission of spirit-beings (both angels and demons) are recognized and stated quite clearly, and that in order to warn us. But nowhere does the Bible descend to the extravagance of fantasia of medieval times, and certainly not to the crudities and excesses of nonbiblical stories. The Bible is not interested in underscoring the form and makeup of spirit-bodies, nor in discussing the permanency of non-permanency of such bodies. What the Bible does say is that these entities belong to the category of spirit-beings, suggesting that they are immaterial and incorporeal in nature. As Hebrews 1:14 states, they are "ministering spirits."

Evidence of Scripture

However, the Bible does declare unequivocally that these spirit-beings can and do assume visibility at will in a bodily form. What is more, this "body" frequently resembles a human. Indeed, at times angels have been mistaken for men. This is why the Bible admonishes, "Be not forgetful to entertain strangers, for thereby some have entertained angels unawares" (Hebrews 13:2).

Abraham was visited by three *men* in the plains of Mamre (Genesis 18:1–8). They walked, talked, sat and ate, just like any other man. But actually they were not human at all but rather spirit-beings from Heaven. Two of the three later visited Lot in Sodom and spent the night at his home. What happened later is one of the most explicit references in Scripture of angels having power to assume solid matter.

Because of the depravity and degradation of the city of Sodom, God decided to destroy it. The city had become a veritable cesspool of iniquity, with its stench reaching to the highest heavens. It still remains a mystery as to how Lot, Abraham's nephew, could have become one of Sodom's leading citizens! Among its vile practices was sodomy—a name directly derived from that of the city itself. When the two spirit-beings arrived in town, the men of Sodom wanted sex with them. Here was new and strange flesh. They called on Lot and demanded that he produce his two guests. "Bring them out to us, **that we may know them.**" Shades of Genesis 6? Could the men of Sodom have known Lot's guests were extraterrestrial? Jude seems to suggest so. "Even as Sodom and Gomorrah, and the cities about them in like manner, giving themselves over to fornication, and going after strange flesh…" The implication is that their sin was not only homosexual but with extraterrestrials in the bargain.

Not only can angels assume solid matter, but they can return also to their non-solid state. Beings who possess this power are known as paraphysical. Angels and demons seem to belong to this category.

Humanoids share the identical characteristic. They have power to assume solid, physical form, and by so doing become visible to humans; they also have the power to become invisible again and take on an ethereal existence.

How Angels Materialize

If angels, demons, and humanoids possess this power to materialize and take on a bodily form, the question that remains is: What sort of body do they have? Can it be analyzed as to its composition? What ingredients are present that a being who is invisible one minute becomes visible the next?

Some have speculated on how this happens but not always with success and certainly not with unanimity. Matter, according to scientists, has an electron-proton-neutron energy base. Millions of electrons spin in their orbits around a tiny pinpoint of energy, called the nucleus. In his book *Did Genesis Man Conquer Space?* Dr. Emil Gaverluk explains the nucleus by stating: "It is in effect an ultramicrominaturized solar system filled with awesome power as we know in the atomic age. Each atom is ready to burst with power. Yet, it is subject to the laws of the electromagnetic spectrum and can be measured, observed, manipulated because of it."[169]

Then he presents the predicament: "Can one examine a spirit under an electrobeam microscope? Can a spectroscope reveal a spirit light spectrum? Can a radioscope, with all of its marvelous electronic capability, listen in to spirit communications? Can a telescope with an image intensifier see spirits flitting across the reaches of space? Of course not. Spirits are not made of atoms. They are structured otherwise."

So much for pure spirits. But what happens when these same spirits assume bodily form? How is the transformation accomplished? With what body do they come? John Keel has attempted to explain the process: "In order to materialize and take on a definite form, these entities seem to require a source of energy; a fire or a living thing—plant, a tree, a human medium (or contactee). Our sciences have not reached a point where they can offer us any kind of working hypothesis for this process. But we can speculate that these beings need living energy which they can reconstruct into physical form. Perhaps that is why dogs and animals tend to vanish in flap areas. Perhaps the living cells of those animals are somehow used by the ultraterrestrials to create forms which we can see and sense with our limited perceptions."[170]

Dr. Clifford Wilson suggests a similar process and so does Kenneth C. Bayman. These beings from space are able to absorb vital energy, vampire-like, from their human dupes and victims, also from animals. This would explain why animals have been killed, mutilated and stolen by UFOs. Vital energy is associated with living creatures and because of this is different from atomic or nuclear energy.

What is fascinating and adding to the mystery is that these modern explanations are not all that removed from that of old Adam Tanner describing the makeup of demonic entities *back in 1629.* "They often form for themselves bodies from impure air or vapors and exhalations or clouds mixed with air. To the air water is added, earth, mud, sulphur, resin, wood. Sometimes too there are added bones from the corpses of animals or condemned men, at times too from the semen of beasts and men and such like matter."[171]

The Wild Man of Gadara

Dr. Wilson, who combines his UFO research with a Christian testimony, finds many of his companions in the New Testament. He suggests that the same physical life principal as already described is found in the incident where Jesus healed the wild man of Gadara. This man was possessed of a "legion" of demons, and ran amok among the tombstones, exhibiting superhuman strength. When Christ appeared on the scene, the demons recognized the presence of a superior power, and sought permission to enter into a herd of swine instead. This is known as "occult transference." The request of the demons was granted, and the New Testament records the herd of swine raced headlong down a steep slope and drowned in a lake. Wilson adds: "Almost as though they knew they were occupied by totally objectionable creatures." He then proceeds to give his own personal belief: "It would seem entirely possible there is a limitation to the powers of these beings, and that the only new life available to them is whatever they can 'borrow,' whether it be human, animal or even energy forces such as electricity."

It would seem that in order to materialize and become visible these spirit-beings have to secure a source of energy, and this would explain many of the strange incidents that are frequently associated with Close Encounters of the Third Kind.

Spirit Séances

As would be expected, this explanation bears a strong resemblance to certain materializations experienced in spiritualist séances, with the aid of a medium. Admitting that some of these materializations have been frauds, there still remains a sizable remnant which defy such categorizing. Some of these materializations have actually been proved genuine, and even photographed. A typical example of such materialization is one quoted by Martensen Larsen. He tells of twelve sessions held with the medium Mm. D'Esperance. "While the medium sat in the 'cabinet,' a phantom built itself upon the floor outside the cabinet and finally took shape as a female person, who moved in and out among the participants. She extended her hand to one of these, and while he held the hand dematerialization set in before the eyes of all. The gentleman, a well-known person, called out, "Now the hand is growing smaller and smaller. Now there is nothing left!" Finally there was only a small ball left on the floor, which rolled into the cabinet."[172] Dr. Kurt Koch speaks of six degrees of teleplastic morphogenesis:

> **The first stage** is the emission and separation of veil-like, gauze-like, slimy substances of rubbery elasticity from the body cavities of the medium...**The second stage** is the formation of body parts such as outline, arms, legs, head, etc....**The third stage** is the solidification into complete shadowy forms, which appear as phantoms near the medium...In these three stages we have purely visual materialization phenomena. In the next three stages we move on to active and passive manifestations of energy by the phenomenon.
>
> **The fourth stage** shows a combination of materialization phenomena with telekinesis. The medium is in a position to display energy at a distance by means of an unknown remote power...**The fifth stage** of the materialization is the penetration of matter...**The sixth stage** of materialization probably makes greater demands on the imagination of a person of sound mind than any other phenomenon in the whole field of parapsychology. This is the occasional testimony given to metamorphosis into animal shapes.[173]

It would seem that the energy source required for these materializations may not be totally different from the energy source used by space-beings when they appear visible on earth. Seemingly, this is about as far as the paraphysical or parascientific theories are able to take us. Some of the phenomena can be explained, but all agree that a vast residue continues to remain as an inexplicable mystery.

When Science Is Silent

Faced with this apparent inability of science, is there anything further that one can do? When human methodology proves inadequate and is unable to supply any concrete solution, is there any other avenue of information that is still open for investigation? When science collapses in face of the nonmaterial and irrational, what more can be done?

It is our belief that when human knowledge reaches the end of its tether, one should seek in the transcendent world for an answer. When reason becomes inadequate, one turns to revelation. When there is no word from man, it is wise to enquire; "Is there a word from the Lord?" Does divine revelation speak when all human experimentation and investigation have become silent? Does God's Word to man, the Bible, have light to shed on this question?

It is significant that Vallee has come to a somewhat similar conclusion: "Scientific data alone will not give us the answer to the UFO mystery."

And then he suggests that one should open up new areas of investigation, "notably the investigation of the religious experience."[174]

Such an approach, of course, is not without precedence. Many times in the past, divine revelation has supplied the answer, which human knowledge had been unable to attain. In some cases, science did catch up a few centuries later and confirmed what God's Word had proclaimed from the beginning.

All this is to be expected, since the Bible has an Author who has omniscience at His disposal. The Bible is the vehicle of communication by which the Creator of the galaxies has made known His thoughts to finite man on planet Earth. After all, if God be God, it is only to be expected that He possesses and reveals data and reality far beyond the realm of empirical science. This is one of the reasons why the Christian believer accepts the Bible as a book that is totally different from any other. It is not a mere compendium of human knowledge containing the scholastic contributions of learned authors. Rather it is written record of God's revelation to man. That is why it uncovers data and truths that human minds would never have succeeded in discovering on their own. It is not that men thought these things but that God declared them. They are not the conclusions and deductions of rational, human minds, but the communications and revelations of the Divine Mind of God.

This is invariably so in matters concerning our faith. Human reason on its own would never have discovered the most elementary and foundational precepts of the Christian faith. It was not man's reason that told him that God is his Father, that God is love, that God has prepared a place for him in heaven...This kind of data comes through divine revelation.

And the same could be claimed of data relative to the physical universe. Human science has discovered many of these facts and by so doing confirmed the revelation found in God's Book. This being so, one is justified in searching deeper into this revelation to discover if it has anything to say on this matter of spirit-beings from outer space? If our beliefs cannot be based on reason, can they be based on revelation? When human science is silent, does divine revelation speak?

12

Revelation to the Rescue

__Devil__ is the opposite of __angel__ only as
__bad man__ is the opposite of __good man.__
Satan, the leader or dictator of devils,
is the opposite, not of God, but of Michael.
—C. S. Lewis

The Bible assumes the existence of a spirit world and of spirit-beings. Introduced early in the Scriptures, the reality of angels and demons is taken for granted. They are depicted as possessing energy and power to an extent denied to mere humans.

But does divine revelation supply more than the fact of their existence and power? Does it explain and interpret the fact? Does it tell who these spirit-beings are? What they are? Whence they come? Whither they go? What is the purpose of their existence? We believe it does.

We are told that they were created by God before the first man ever appeared upon the Earth. In other words, spirit-beings have been around for a long time. They are not, however, eternal or everlasting. At some point in time they were created by God, and at a later point in time many pitted their will against God. Back in the primeval world these spirit-beings rebelled against their Maker, but the coup was unsuccessful. They succeeded only in bringing judgement upon themselves. We are not given a direct, specific account of this event, but it is a reasonable deduction from the evidence available in Scripture. The end result of it all was that these sprit-beings were dismissed from the presence of God and emerged as demons or evil spirits. Originally they were angels in Heaven, but after allying themselves with Lucifer in his rebellion, they became fallen-angels or demons. They are the ones responsible, with Lucifer, for introducing evil into God's perfect world.

Introducing Lucifer

Now that we have introduced his name, who exactly is Lucifer? He is certainly not the silly caricature portrayed by poets and artists for the Middle Ages, nor the grotesque creature painted in such lurid colors, with horns, hooves, a forked tail and red tights —often carrying a large trident and sometimes a pair of bat wings for extra effect! To most people the devil became so ridiculous a figure that he was laughed out of existence.

Two hideous horns on his head he had got,
Like iron heated nine times red-hot;
The breath of his nostrils was brimstone blue,
And his tail like a fiery serpent grew.[175]

This was a tactic of masterly cunning on the part of the devil himself. The fact that a creature who was at one time an archangel in Heaven, superior and brilliant, the very epitome of wisdom, could now be depicted as a stupid, grotesque animal, only proved his effective ingenuity in the manipulation of human minds.

When it comes to a portrait of the devil, Milton is far closer to the mark than Southey.

> The other shape
> If shape it might be called, that shape had none
> Distinguishable member, joint, or limb
> Or substance might be called that shadow seemed,
> For each seemed either; black it stood as night
> Fierce as ten furies terrible as hell.[176]

What is true of the devil applies equally to the angels who followed him. Like their leader, they possess great knowledge but are not omniscient. They move with incredible swiftness from one realm to another but are not omnipresent. They have much power and authority but are not omnipotent. Mighty but not all-mighty.

Some believe that demonic beings have unlimited power. This theory has infiltrated branches of the modern church, but it is contrary to Scripture.

Power from Below

Rousas John Rushdoony in a penetrating article entitled, "Power from Below" reminds us of this modern image of Satan:

> The new image of Satan is a product of Darwin and Freud. **First,** he is not a creature made by God but a dark force evolved out of chaos and essentially is chaos. There was never thus any higher status for Satan, but a totally subterranean one, a creature of chaos, not of God! **Second,** the new image of Satan is of a totally mindless, irrational, perverse being, whose existence is total terror to the rationale in man. The new Satan is the utter contradiction of reason, whereas the biblical Satan is the example of a fallen and totally depraved person. It should not surprise us that some of the new Satanists lose their reason and become themselves mindless. **Third,** because no

> power exists above by definition, total power is held to exist below, and the result is a growing evidence that there is a strong tendency to believe in the omnipresence and even omnipotence of Satan. Satan is held to be everywhere, operative in all situations, and hence to be reckoned with at all times. It is surprising how far this idea has infiltrated the churches. Too few churchmen remember that Satan, like themselves, is a creature, capable only of a local appearance, i.e., able to be in only one place at a time. Only God is omnipresent and omnipotent.[177]

Although short on omnipotency, demons nonetheless do exercise enormous power and influence, far exceeding that or ordinary mortals. They are frequently designated in the Bible as "principalities and powers." The kingdoms of the world belong to them, and their leader Satan is addressed as "the prince of this world." As C. S. Lewis reminds us: "There is no neutral ground in the universe: every square inch, every split second is claimed by God and counter-claimed by Satan."[178]

Part of his power lies in the fact that he is able to wreak untold havoc on individuals, and indeed on nations. Who can deny the influence he wields over certain nations of our world today? And over some of the despots who govern them? Who can deny that many a tyrant was demonically manipulated and issued decrees that resulted in the slaughter of millions?

Threatening the Royal Seed

But even these are tangential compared with the main mission of these demonic beings. Their chief purpose is nothing less than to thwart God's entire plan of redemption. This has been their major target from the beginning: to destroy the Royal Seed and so deprive man of his rightful place in God's Kingdom.

Following their initial rebellion, certain of these "fallen angels" fell again the second time in the event described in Genesis 6. They are the ones who lusted after the women of Earth, married them, and produced a prodigious offering known as the Nephilim. This was another link in the chain of events to destroy the Royal Seed. This time it was to be done by contaminating the human race.

Because of the particular villainy of this event, God imprisoned these angels in Tartarus as His own special prisoners. The other "fallen angels" remain temporarily free, using their freedom to continue their nefarious activities.

The Omega Conspiracy

As for the future, the Bible predicts that before Christ's return to Earth, these same spirit-beings will engage in one final, titanic effort to again seduce the human race. Their human agent is called the Antichrist, whose "coming is after the working of Satan" (2 Thessalonians 2:19). This is *The Omega Conspiracy*, dwarfing every previous attempt to destroy mankind, and by so doing, depriving Christ of the victory that rightfully belongs to Him.

When Christ was on Earth the first time, He asked His followers: "When the Son of Man cometh, shall he find faith on the earth?" (Luke 18:8). Demonic entities are determined that the answer to that question be a resounding "no." It is their mission that faith disappear completely from the planet.

Are They Back?

Is it feasible that those entering our atmosphere today, under the guise of space-beings and traveling in mysterious space-vehicles, are these same demonic agencies? Is this the first phase of their planned return to conquer the Earth?

Jacques Vallee does not use biblical terminology, nor does he write from the standpoint of Scripture; nonetheless, he comes surprisingly close to the conclusions of the Bible. He writes:

> The experience of a close encounter with a UFO is a shattering physical and mental ordeal. The trauma has effects that go far beyond what the witnesses recall consciously. New types of behavior are conditioned, and new types of beliefs are promoted. The social, political, and religious consequences of the experience are enormous if they are considered, not in the days or weeks following the sighting, but over the timespan of a generation. Could it be that such effects are actually intended, through some process of social conditioning? Could it be that both the believers and the skeptics are being manipulated by what I jokingly call the Higher Intelligence Agency? Is the public being deceived and led to false conclusions by someone who is using the UFO witnesses to propagate revolutionary new ideas?[179]

Denouement

Two thousand years ago, the apostle Paul warned his readers of this very thing. He told them of the future appearance of Antichrist on the world scene and how his activities would be controlled and masterminded by Satan himself. Paul elaborated by saying that Antichrist's coming would be characterized by "all kinds of counterfeit miracles, signs, and wonders, and in every sort of evil that deceives those who are perishing" (2 Thessalonians 2:9–10, NIV). This is the "social conditioning" in which these alien and demonic beings are engaged. It is to delude and deceive the human race.

In a similar vein Christ had warned His followers: "If anyone says to you, 'Look, here is the Christ!' or, 'Look, there he is!' do not believe it. For false Christs and false prophets will appear and perform signs and miracles to deceive the elect—if that were possible" (Mark 13:21–22, NIV).

From their past record, we know that demonic beings have amazing power and are able to achieve impressive results. They have the gift both to appeal and to appall. "Millions already have come to their harm, / Succumbing like doves to his adder's charm."[180]

All this is but a foretaste of what lies ahead when these demonic hordes are unleashed in their full fury upon our planet. And adding to the grim reality is the possibility that it could happen in our lifetime. The biblical signs of the end are already being fulfilled, and this ultimate sign of demonic agencies trying to control and seduce the human race may be just around the corner.

For their conspiracy to succeed, these agencies know they will have to be extremely cunning. They will have to attack with "limitless deceit" (2 Thessalonians 2:10, Berkely Version). They will pose as our friends and allies, sent here to rescue us. Humans will experience the "wiles of the devil" as never before. Demonic attacks will be carried on with such subtlety and deception that God has seen fit to warn us well in advance: Watch out that no one deceives you" (Matthew 24:4, NIV). It will be deceit and deception such as the world has never seen. Dave Hunt and T. A. McMahon attach such importance to this fact that they state that "the consequences of being deceived...are far worse than being victimized by famine, disease or war."[181] We are repeatedly warned in the Bible to be informed on this matter and not to be ignorant. There is no other defense against demonic deception but a thorough acquaintance with the truth.

Jesse Penn Lewis makes a profound statement when he says: "Since the deception is based on ignorance, and not on the moral character, a Christian who is 'true' and 'faithful' up to the knowledge he has must be open to deception in the sphere where he is ignorant of the 'devices' of the devil (2 Corinthians 2:11) and what he is able to do. A 'true' and 'faithful' Christian is liable to be deceived by the devil because of his **ignorance.**"[182]

The devil's ultimate target is to bring man under his control, and the only way he can achieve that is by gaining entry into man's mind. It is there that the battle is joined. The devil is out to confuse the mind (2 Thessalonians 2:2), to divert the mind (James 4:8), and to discourage the mind (Hebrews 12:3).

Ignatius reminds the Ephesians in a letter he sent them: the devil "bewitches his victims before he destroys them." Tertullian also was equally conversant with the devil's motives: they were such that "men may be willing to know for certain, what they certainly do not know."[183]

With the accumulation of the end signs, it is within the realm of possibility—if not probability—that today's generation may actually see the return of the Nephilim. If nothing else can shock our sophisticated society from its sinful stupor, a demonic invasion from space could be the one catalyst to do so.

The Conspiracy Doomed

The practical question that remains is: how can finite man withstand, let alone repel, such an invasion? How can frail "flesh and blood" stand up to these superior intelligences from outer space? If the archangel Michael dared not to withstand Satan, how can ordinary mortals, created so much lower than the angels?

The answer, of course, is *Christ.* His name and His authority are the only resources that are adequate to repel such entities. Proof of this power is supplied in abundance in the New Testament. On page after page Christ commands these entities to depart, and each time they obey. Let that be repeated: *each time they obey.* There is no recorded exception! Christ's own enemies were the first to admit such power: "For with authority commandeth He even the unclean spirits, and they do obey Him" (Mark 1:27).

The wonder of His power is not limited to His ability to cast out demons but rather to do so without exception and without fail. Other exorcists of the day frequently failed, but as for Christ there is not a single instance where He was not instantly obeyed.

It also surprised everyone in Jesus' day that He was able to exorcise demons without an elaborate ritual or pseudomedical formula. Christ simply gave the command, and it was done.

Similarly, it is in the name of Jesus Christ and by the power of His Spirit that we can exercise the same authority. God's Word is the guarantor of that promise. No demonic force can prosper against us, As for the begotten of God, the "wicked one toucheth him not" (1 John 5:18). It was because of this that Martin Luther was enabled to sing with such confidence:

> And though the world with devils filled
> Should threaten to undo us,
> We will not fear for God has willed
> His truth to triumph through us.

It is an awesome scenario, but our generation may witness a greater incursion of demonic beings than was experienced in the antediluvian world of Genesis 6. And if they are coming, and integral part of their conspiracy is to mate with modern "daughters of men" and produce a similar hybrid offspring.

In God's Word we have been forewarned of these events and forearmed to withstand them. Although strange and frightening things are happening on Earth in these end-times, we are not left alone for the battle. Our focus is not to be on the return of the Nephilim but rather on the return of Jesus Christ, the Alpha and Omega, the King returning to reign. He is the mighty, triumphant Conqueror by whom every foe, terrestrial and extraterrestrial, is ultimately vanquished. Satan's final conspiracy against Christ and His Kingdom will not succeed. At no time is the outcome in question. This we know, because we have advance information on how the conflict is going to end. That knowledge is given to us in God's Word, and it carries the divine imprimatur. We are even given a preview of tomorrow's headlines: "The kingdoms of this world are become the kingdoms of our Lord, and His Christ; and He shall reign for ever and ever" (Revelation 11:15).

13

Satan and Sex

Assuming that the identification can be made—that demons from hell and humanoids from space are one and the same—is there any evidence of their interest in sex?

This was the crucial point in Genesis 6. The "sons of God" *married* the "daughters of men" and *had offspring from them.* Is there any other documentation that would link the demonic world with this very human trait of sexual desire?

Hollywood seems to reflect very adequately the seamy and sexual side of American society. It has known for many years that illicit sex is its best-selling merchandise. Because of this she has churned out an avalanche of movies majoring on obscenity, vulgarity, sodomy, voyeurism, transvestity, bestiality…

But what of sex films that include the demonic world within their purview? Do Satan and sex go together?

Hollywood may think that it has said it all and covered every sin in the catalog. But actually, Hollywood has not said it all. The savants of the celluloid empire have still to portray the ultimate. And when that sin is committed it will be viler than fornication, adultery, sodomy, bestiality…or all these put together. It is carnal relations with extraterrestrial beings. Some of Hollywood's productions have got close to it but the ultimate is still to come. This will be sexual union between women of earth and demons of hell.

A natural response is to gasp and say: "Impossible! Such a thing can never happen!" The very premise of this chapter is to say: "It has happened!" Genesis 6 supplies the documentation.

This happened back a long time ago in the days of Noah, but what is of frightening import to our generation is that the New Testament predicts that some of the distinctives of the days of Noah will be repeated at the end-time. "And as it was in the days of Noe, so shall it be also in the days of the Son of man" (Luke 17:26, King James Version).

But why should Satan be interested in sex? In what way can he advance his cause by the intermingling of demons and humans in sexual relations? One reason may be found in his desire to downgrade woman. He remembers that Christ came as "the seed of the woman." Another reason is his desire to "get back" at God. As Dr. Hymers states in his manuscript, *Sex and Prophecy*, "Since he cannot hurt God Himself, he attacks men who God loves. Thus he indirectly attacks God, And one of the devil's most powerful weapons against humanity, is to twist and pervert man sexually."

Knowing that his time is short, he accelerates his activity. "...The devil is come down unto you, having great wrath, because he knoweth that he hath but a short time" (Rev. 12:12).

But there is another reason: the devil uses sex and orgasm to implant his ideas into his victim's mind. He is aware that one of the human experiences that comes closest to the ecstasy of communication with God is that of sexual orgasm between lovers. If ever a human being acts as if he were possessed without actually being possessed, it is when he is overwhelmed by the irresistible tide of desire. Here he trembles, groans, writhes, cries out...in a bittersweet experience that defies description.

William Sargant points out another similarity: "It is significant that 'having,' 'knowing,' and 'possessing' are among our commonest expressions for sexual intercourse, for they suggest that the real goal and summit of sexual activity is not the procreation of children, or even erotic pleasure, but rather the sense of mingled identity that lovers briefly achieve, the acquisition of another human being who, if only momentarily, seems to become part of oneself…Numerous Christian mystics have described the human soul as the female, surrendering to and possessed by God as the male."[184]

This may be the reason why German mystic Heinrich Suso referred to himself throughout his autobiography as "she."

Manipulating Sex

Satan used and manipulated sex to give man a simulated spiritual ecstasy. But rather than the ecstasy being divine, his version has been demonic. In this way Satan has succeeded in introducing his thoughts and ideas into man's being, under cover of an erotic ravishment. And by the same method he has been able to confirm and consolidate his work.

Donald Nugent in "Renaissance and Rise of Witchcraft" says: "There seems to be certain common denominators found rather universally in witchcraft. It particularly orients around two things: sexuality…and power, often fusing the two."[185]

Kent Philpott illustrates this from a personal counseling session:

> A twenty-eight-year-old junior college student came to see me, for he had heard that I knew something about demon possession. He had worshipped Satan with Anton La Vey's group in San Francisco and became a member of the Church of Satan, having sold his soul to the devil. In exchange for his soul, he was to receive power to make money. It worked – he made lots of money with very little effort. Then he made a switch. Instead of power to make money, he wanted power over women. He received that, too. When he met a woman he

wanted, he would concentrate on her. Before the day was out, the woman would either show up at his door or call him on the phone. He came to me because the women told him that he was strange. Several told him he was a demonic. After we'd talked a long while, he reached a conclusion: he didn't want Jesus because, he said, "I'd lose my power."[186]

In most cases, Satan has been satisfied in using eroticism between humans as a vehicle for his devices. However, there are many instances where he has engaged demons more actively in the sexual process. Such demonic enslavement to lustful sex is known as incubacy and succubacy. No aspect of demon possession is more revolting.

Novelists have woven many story plots around satanic intercourse and impregnation. Examples from contemporary literature are *Rosemary's Baby* by Ira Levin, *The Violent Bear It Away* by Flannery O'Conner, *The Devil's Own Dear Son,* by James Branch Cabell, *The Sound and the Fury* by William Faulkner, and *Black Eater* by James Blish. But perhaps the most grotesque description of all is that by Isaac Bashevis Singer in his novel *Satan in Goray.* Here, demonic sex and demonic oppression are laid bare in front of us.[187]

Incubacy and Succubacy

But these are imaginary stories. What of historical facts? Do we have any evidence from history that relates to incubacy and succubacy?

In his classic, *De Civiatate Dei,* Augustine states: "It is a widespread belief that Sylvans and Fauns, commonly called incubi, have frequently molested women, sought and obtained coitus from them."

Witch trials frequently supplied testimony from women alleging they had sexual intercourse with incubi. And according to W. Raymond Drake, men also "described delightful nights with beautiful succubae."[188]

It is common knowledge that Tantric cults of the East nave used sexual intercourse in order to induce possession by demonic powers. This is not exactly the same as incubacy and succubacy, but the end result is frequently similar.

Orleans, France 1002

There is evidence from France, from the eleventh century, of this association of sex and Satanism. In 1002 in the town of Orleans, a number of heretics were put on trial at Christmas time. Along with heretical doctrines the accused were also charged with something that was quite different in character:

> Gathered indeed on certain nights in a designated house, everyone carrying a light in his hands, and like merry-makers they chanted the names of demons until suddenly they saw descended among them a demon, in the likeness of some sort of little beast. As soon as the apparition was visible to everyone, all the lights were forthwith extinguished, and each with the least possible delay, seized the woman who first came to hand, to abuse her, without thought of sin. Whether it were a mother, sister, or nun whom they embraced, they deemed it an act of sanctity and piety to lie with her. When a child was born of this most filthy union, on the eighth day thereafter a great fire was kindled and the child was purified by fire in the manner of the old pagans, and so cremated. Its ashes were collected and preserved with as great veneration as Christian reverence is wont to guard the body of Christ, being given to the sick as a viaticum at the moment of their departing this world...[189]

The Black Mass

Later in the same century, again from France, the Black Mass was devised and recited. Composed as a parody of the Roman Catholic

Mass, it was celebrated to cause harm to one's enemies, and to spread the influence of evil. For the bread and wine of the holy sacrament they substituted such vileness as a toad, a turnip, a mixture of blood, urine, and feces. It was alleged that little babies were sometimes sacrificed in the ritual. And, of course, sex acts were committed. It is claimed that Madame De Montespan, mistress of Louis XIV, offered herself sexually as a living sacrifice in order to retain the king's love. With the passage of time, sex orgies became commonplace, and all sorts of radical, hedonistic practices were introduced.

In the eighteenth century, Hell Fire Clubs sprang up all over Europe, counting among their patrons some of the leading aristocrats of the day. In one such club, Medmenham Abbey in England, membership included several earls and the Chancellor of the Exchequer.

However improbable these happenings may have seemed to some people, they should cause no problem to us today. With the resurgence of Satanism and the appearance of cults like that of the Manson family, indulging in promiscuous orgies and ritualistic murders, their credibility can no longer be questioned.

San Francisco 1966

The Church of Satan in San Francisco is a typical example. Anton Szandor LaVey established this "church" in 1966 after receiving his demonic empowering on the night known as Walpurgisnacht—the most important festival in the lore of magic and witchcraft. This "church" is reputed to have up to ten thousand members. In 1968 LaVey published his "Satanic Bible," a parody on the Holy Scriptures. Here are some "choice" verses:

> Blessed are the powerful, for they shall be revenged
> among men.
> Cursed are the feeble, for they shall be blotted out!
> Blessed are the bold, for they shall master the world!

> Cursed are the righteously humble, for they shall be trodden down under cloven hooves.

A thorough-going extrovert, he had the word VAMPYR imprinted on his automobile license plate! He even kept a lion as a household pet! LaVey's "satanic" weddings have gained him much notoriety, and so has his ritual of using a nude girl as an altar.

All over the world, including American cities, there are "churches" dedicated to the worship of his "Infernal Majesty," Satan. Such devil worship involves actual communication with evil spirits and with "principalities and powers." Here the whole atmosphere is charged with terror and nausea.

> Often the devotees use a ruined or deserted church, sometimes an overgrown cemetery. Their ritual begins at eleven o'clock at night, aiming to finish at twelve. Central to the ceremony is an altar with black cloth, six black candles, a chalice, and crucifix turned upside down. The altar is a naked woman lying on a table and holding black candles in her hands. The priest consecrates the host on her bare stomach. The Black Mass is the Roman Catholic Mass verbatim, with the word Satan substituted for Christ. The Roman Catholic host is often stolen from the real Mass in a handkerchief. The high point of the ritual comes when the priest has intercourse with the girl on the altar, and act often accompanied by crude sadism. Many such evenings end with drunken dancing, drug taking, and a general sex orgy; often the next day there are found skinned cats, beheaded chickens, or ritual bags containing drugs, potions, animal bones, and occasionally human fingers.[190]

In close association with rites of these kinds are the bizarre and demonic cults that have reappeared in our Western societies in the last few decades. There can be no mistake about it, when one falls prey to demonism, he has reached the ultimate in occult practice. Here the color is no longer grey but rather midnight black.

Is There Biblical Evidence?

Evidence abounds in legends, literature, and actual history that Satan uses orgasm for his nefarious purposes, that demons and humans come together sexually. One question remains: Beyond Genesis 6, is there any evidence in the Bible of such things taking place? Yes, in Isaiah 34:14: "The wild beasts of the desert shall also meet with the wild beasts of the island, and the satyr shall cry to his fellow; the screech of the owl also shall rest there..."

On the surface the evidence seems obscure. The key to interpreting this passage is "screech of the owl." The original Hebrew word is *Lilith*—a word that has caused no end of difficulty in translation. Many modern versions translate the word as "night creatures," but this is equally obscure. Most scholars now believe that the word is borrowed from the Assyrian and refers to the *female demon of the night.* In Assyrian mythology lilith was a female demon and succuba and had intercourse with men in their dreams.

Sodomy and Sodom

One should also read Genesis 19. Here one finds a more direct reference to sex relations between spirit-beings and human beings. Two spirit-beings from Heaven visited Lot in the city of Sodom and warned him of the impending doom of the city. The depravity and degradation of Sodom was such that God decided to destroy it. The city had become a veritable cesspool of iniquity, with its stench reaching to the highest heavens. It still remains a mystery as to how Lot, Abraham's nephew, could have become one of Sodom's leading citizens and yet be called "righteous Lot." But if he was righteous, it is obvious the other citizens were not. Among their vile practices was sodomy—a term directly derived from the name of this city. When the two spirit-beings or angels arrived in town, the men of Sodom wanted sex with them. Here was new and strange flesh. "And they called unto Lot, and said unto him, Where are the men which came into thee this night? Bring them out unto us, **that we may know them**" (Gen. 19:5).

Shades of Genesis 6? Could the men of Sodom have known that Lot's guests were extraterrestrial? Jude 7 seems to suggest so. "Even as Sodom and Gomorrah, and the cities about them in like manner, giving themselves over to fornication, and going after strange flesh..." The implication is that their sin was not only homosexual but with extraterrestrials in the bargain.

Humanoid-Human Sex?

Many contemporary reports of Ufology approximate the event in Genesis 6. In hundreds of reported cases, humans have been contacted by beings from space, and scores have been taken aboard space vehicles. Some of these incidents of sexual assault have been documented. Whether there is evidence of offspring as a result of those assaults is debatable. There is cause to believe in humanoid-human sexual encounters, but we cannot say for sure that births of later-day Nephilim have occurred. But the few cases of humanoid-human unions that have been reported come close, uncomfortably close.

Marlene Travers of Melbourne, Australia, is an example. She claims that she was raped by a tall handsome man from a UFO, on the night of August 11, 1966. When doctors examined her they found burns on her arms and legs—and that she was pregnant.[191]

A California woman reported a similar instance of rape by a spaceman and gave birth to an unusual but stillborn baby.[192]

Probably the best documented case of sexual assault is that of Antonio Villas Boas from Brazil.[193] This is an incident that deserves every attention and scrutiny. It occurred on the night of October 15,

1957—a bizarre happening if ever there was one. In this instance it was not a man from space having sex with a woman from Earth, but the reverse.

How well documented is this case? Here we are indebted to Dr. Olavo Fontes, a physician, and Senor Joao Martins, a newspaper man, for the reports that have been made available. Both these men, it should be noted, were skeptical of the story when they first heard it. They both believed it too fantastic to merit credibility.

Antonio Villas Boas was a twenty-three-year-old farmer of Sao Francisco de Sales, who claimed he was forcibly taken aboard a flying saucer, and was made to have sexual intercourse with a female aboard the craft.

That night of October 15, 1957, Villas Boas was plowing with his tractor, when he saw a large, luminous, egg-shaped object flying toward him at terrific speed. Suddenly the object halted and descended until it was some fifty meters above his head. All at once it lit up the tractor and all the ground around, as though it were daylight. The pale, red glare of the vehicle was so powerful that the tractor lights were swamped by it. Slowly the craft dropped toward the ground, until three metal supports—forming a tripod—emerged from beneath to take its weight.

Panic stricken, Villas Boas started up his tractor to make his escape. All at once the engine died, and simultaneously the tractor lights went out. Try as he would he could not restart the engine. He then jumped to the ground and began to run as fast as he could.

He was overtaken by four individuals and carried aboard the machine.

After describing in some detail what he saw aboard the craft, Villas Boas comes to the central feature of his story. This included a very detailed description of the woman. It also included details of what took place between them which are too sexually explicit to print here.

In all, Villas Boas had spent four and a half hours aboard the craft. When he was finally released and helped out of the vehicle, he made for his home as fast as possible. He was too ashamed, however, to tell his story to anyone, that is, to anyone except his mother. It was many weeks later that he contacted the newspaper journalist Jaoa Martins.

After some reluctance on the part of Villas Boas, the newspaperman was able to persuade him to relate his experience to himself and to the medical doctor Olavo Fontes. For four hours they listened to the incredible tale. They subjected him to an exhaustive interrogation and cross-examination, but young Villas Boas stuck to his story throughout, strange as it seemed.

Following this he was given medical and psychological tests, and it was discovered that yellow spots and sores had broken out on his skin, and that these seemed to indicate radiation exposure.

People who further investigated the testimony of this small-time farmer from the Brazilian outbacks have reluctantly concluded that he was telling the truth.

This whole subject is unfamiliar territory to most of us. The theme is so strange and grotesque that it takes a quantum leap of faith to accept it. But such things have happened in the past with biblical evidence to endorse them. It seems as we approach the end-time, we are beginning to see a recurrence in phenomena that baffles finite minds. Could the return of the Nephilim be the next item on the agenda?

About the Author

I. D. E. Thomas is one of a long line of Welsh preachers, and he has traveled extensively in Europe, the Orient, and on the American continent. His sermons have been described in the press as "powerful and passionate," "intellectually articulate" and "spiritually probing."

A native of Wales in the United Kingdom, Dr. Thomas has held three pastorates in that country: the Amman Valley, Caernarvon and Llanelli, where he succeeded the Reverend Jubillee Young.

For ten years he conducted special preaching missions in the United Kingdom and the United States. He also served as Commentator for the BBC (British Broadcasting Corporation).

Notes

1 Edwyn Bevan, *Symbolism and Belief* (Boston: Beacon Press, 1957), 48.

2 Alan and Sally Lansburg, *In Search of Ancient Mysteries* (New York: Bantam Books, 1974), 87.

3 Ibid., 87.

4 Ibid., 38.

5 Andrew Thomas, *We are not the First* (New York: Bantam Books, Inc., 1973), 3.

6 Alan and Sally Lansburg, *In Search of Ancient Mysteries* (New York: Bantam Books, 1974), 111-112.

7 Christopher Lasch, *Time Magazine,* March 5, 1979.

8 R. L. Hymers, *Encounters of the Fourth Kind* (Van Nuys, California: Bible Voice Inc., 1976), 47.

9 Zbigniew Brezinski, *Between Two Ages* (New York: Penguin Books, 1976), 56-57.

10 Dr. Gordon J. F. MacDonald, *How to Wreck the Environment (Geophysical Warfare), Unless Peace Comes,* ed. Nigel Calder (Viking Press, 1968).

11 Quoted in *The Los Angeles Times,* March 3, 1984.

12 Cited by Woodrow Nichols and Brooks Alexander in the *SCP Journal* (1977), under the title "The Modern Promotheus," 5.

13 *SCP Journal*, 6.

14 *SCP Journal*, 6.

15 John A. Keel, *Operation Trojan Horse* (London: Sphere Books Ltd. 1973).

16 Marcia Seligson, *New West Magazine* (November 7, 1977).

17 Graffiti found in 1967…Vide: Ralph Blum with Judy Blum, *Beyond Earth* (New York: Bantam Books, 1974), 155.

18 Psychotronics—similar to parapsychology

19 Jaques Vallee, *Messengers of Deception* (Berkeley: And/Or Press, 1979), 146.

20 J. Allen Hynek, *The UFO Experience—A Scientific Inquiry* (Chicago: Henry Regnery Co., 1972).

21 Quoted in the *Los Angeles Times,* November 24, 1979.

22 Dr. Hynek on the *Merv Griffin Show*, January 12, 1978.

23 Quoted by Barry H. Downing, *The Bible and Flying Saucers* (London: Sphere Books Ltd., 1968), 55.

24 Donald A. Keyhoe, *Flying Saucers from Outer Space* (New York: Tandem Publishing Co., Ltd. 1970), 60-62.

25 Clifford Wilson, *UFOs and Their Mission Impossible* (New York: Signet Books 1974), 12.

26 Jaques Vallee, *The Humanoids,* ed. Charles Bowen (Chicago: Henry Regnery Company, 1969), 33.

27 J. Allen Hynek, *The UFO Experience—A Scientific Inquiry* (Chicago: Henry Regnery Company, 1972), 2.

28 *Newsweek Magazine,* November 21, 1977, 97.

29 Preface, VII.

30 John Weldon, *What on Earth is Happening* (Eugene, Oregon: Harvest House Publishers, 1975), 72.

31 Brad Steiger and Joan Whritnow, *Flying Saucers are Hostile* (New York: Universal Publishing & Distributing Corp., 1967), 81.

32 Robert Barry in the *Christian Beacon,* December 21, 1978.

33 Mark Albrech and Brooks Alexander, *SCP Journal* (1977): 14.
34 Otto Binder, *Unsolved Mysteries of the Past* (New York: Tower Publications, Inc.), 61.
35 Mark Albrech and Brooks Alexander, *SCP Journal* (1977).
36 Kelly L. Segraves, *Sons of God Return* (Old Tappan, N.J.: Spire Books, Fleming H. Revell Co., 1975), 16.
37 Quoted by Albrech and Alexander, 14.
38 Quoted by Donald E. Keyhoe, *Flying Saucers From Outer Space* (New York: Universal Tandem Publishing Co., Ltd., 1970), 35.
39 John Weldon, *What on Earth is Happening* (Eugene, Oregon: Harvest House Publishers, 1975), xi.
40 Allen J. Hynek, *The UFO Experience* (Chicago: Henry Regnery Co., 1972), 138.
41 *National Enquirer.*
42 Allen J. Hynek, *The UFO Experience* (Chicago: Henry Regnery Co., 1972), 139.
43 Ibid., 111.
44 Quoted by Donald Keyhoe in *Flying Saucers from Outer Space* (New York: Universal Tandem Publishing Co., Ltd., 1970), 35.
45 Fred Hoyle, *Of Men and Galaxies* (Seattle: University of Washington Press, 1964), 47.
46 Allen Hynek, *The UFO Experience* (Chicago: Henry Regnery Co., 1972), 233.
47 Donald H. Menzel and Earnest H. Taves, *The UFO Enigma* (Garden City, NY: Doubleday, 1977).
48 Walter Sullivan, *We Are Not Alone* (New York: A signet Book, 1964), 228.
49 Ibid., 231.
50 Carl Sagan, ed., Appendix: *Communications with Extraterrestrial Intelligence* (Massachusetts Institute of Technology Press, 1973), 362-364.
51 Dr. Clifford Wilson, *UFOs and Their Mission Impossible* (New York: Signet Book, 1974), 133.

52 Allen J. Hynek, *Edge of Reality* (Chicago: Henry Regnery Co., 1975), XII-XIII.

53 *SCP Journal* (1977): 17.

54 Jaques Vallee, *Messengers of Deception* (Berkeley: And/Or Press, 1979), 8.

55 Ibid., 209-210.

56 Quoted by Donald Keyhoe, *Flying Saucers from Outer Space* (New York: Universal Tandem Publishing Co., Ltd., 1970), 89.

57 Carl Gustav Jung, *Flying Saucers: A Modern Myth of Things Seen in the Skies.* Translated by R. F. C. Hull (New York: Harcourt and Brace, American Library, 1939).

58 Clifford Wilson, *UFOs and Their Mission Impossible* (New York: Signet Book, 1974), 155-156.

59 John A. Keel, *Operation Trojan Horse* (London: Sphere Books Ltd., 1973), 182.

60 Ibid., 60.

61 Volume 15, Number 6: 23.

62 Jaques Vallee, *The Invisible College* (New York: E. P. Dutton 1975), 202.

63 *Voice Magazine,* (July-August 1978).

64 Alan Landsburg, *In Search of Ancient Mysteries* (New York: Bantam Books, 1974), 185.

65 *Matthew Henry's Commentary* (Grand Rapids: Zondervan Publishing House, 1961).

66 Aurelius Augustine, *The City of God,* Trans. Marcus Dods (Edinburgh: T. & T. Clark, 1949).

67 Lewis Sperry Chafer, *Systematic Theology,* Volume 2. (Dallas: Dallas Seminary Press, 1947), 23.

68 Kenneth S. Wuest, *Word Studies in the Greek N.T.*, vol. 4 (Grand Rapids: Wm. B. Eerdmans Publishing Co., 1966), 240.

69 Ibid., 240.

70 Ibid., 241.

71 Unger, *Biblical Demonology* (Wheaton: Van Kampen Press, 1957), 48.
72 W. F. Allbright, *From the Stone Age to Christianity* (Baltimore: John Hopkins Press, 1940), 226.
73 Bernard J. Bamberger, *Fallen Angels* (Philadelphia: The Jewish Publication Society of America, 1952), 53.
74 Philo, *DeGigantibus*, 58-60.
75 *The Ante-Nicene Fathers*, Vol. 8, 85 and 273.
76 Ibid., 190.
77 Josephus, *The Work of Flavius Josephus; Antiquities of the Jews* (London: G.G. Rutledge), 1.3.1.
78 Joseph Hall, *Contemplations* (Otisville, Michigan: Baptist Book Trust, 1976), 10.
79 *Companion Bible* (Oxford University Press), Appendix 26.
80 *The Gospel Truth* Magazine 18, no. 7 (June 1978).
81 Dr. Morgenstern, *Hebrew Union College Annual*, XIV, 29-40, 114ff.
82 Finis Dake, *Annotated Reference Bible*, 63.
83 Kent Philpott, *A Manual of Demonology and the Occult* (Grand Rapids: Zondervan Publishing House, 1973), 77–78.
84 Sometimes spelled as "Bit-Enosh," i.e., the daughter of Enosh.
85 R. H. C. Charles, *Fragments of the Book of Noah* (London: Oxford University Press).
86 Theodor H. Gaster, *The Dead Sea Scriptures* (Garden City, New York: Doubleday and Co., Inc., 1956), 257.
87 Michel Eyquem de Montaign (1533-92).
88 Bernard J. Bamberger, *Fallen Angels* (Philadelphia: The Jewish Publication Society of America, 1952), 25.
89 Daniel 4:13, 17, 23.
90 Jacob Z. Lauterbach, *Hebrew Union College Annual* IV, 173ff.
91 Comyns Beaumont, *The Riddle of Prehistoric Britain* (London: Rider & Co., Ltd.).

92 Trans. by R. H. C. Charles, (London).
93 Dr. R. H. C. Charles, *The Book of Jubilees* (London: A & C Black, 1902).
94 Theodore H. Gaster, *The Scriptures of the Dead Sea* (London: Secker & Wartburg, 1957).
95 Theodore H. Gaster, *The Scriptures of the Dead Sea* (London: Secker & Wartburg, 1957).
96 Ibid., 31.
97 Andrew Thomas, *We Are Not the First* (New York: Bantam Books, Inc., 1973).
98 Tom Horner, *Sex in the Bible* (Rutland, Vermont: Charles E. Tuttle Co., 1974), 145.
99 Emil Gaverluk, *Did Genesis Man Conquer Space?* (Nashville: Thomas Nelson, Inc., 1974), 89.
100 Erich Von Däniken, *Gods from Outer Space* (New York: Bantam Books, Inc., 1972), 161-162.
101 Ann, XV, 44.
102 Sat. i. 155.
103 Bullinger differentiates between *angel* and *spirit* by stating that one defines the office and the other denotes the nature.
104 Kenneth Wuest, *Word Studies in the Greek New Testament* (Grand Rapids: William Eerdmans Publishing Co., 1966), 98-99.
105 Joseph B. Rotherham, *New Testament Critically Emphasized* (Grand Rapids: Kregel, 1959).
106 John Milton, *Paradise Lost,* 6K iv, 1.73.
107 Ibid., 159-160.
108 Tr. I. Epstein 1.218.
109 Some writers such as Ben Adam believe the word "Nephilim" refers to the fallen angels themselves and not to their offspring.
110 Merill Unger, *Demons in the World Today* (Wheaton: Tyndale House, 1971), 26.

111 Quoted by F. Seth Dryness, Jr., *The Journal of Christian Reconstruction* 1, no. 2 (Winter 1974): 51.

112 Kurt Koch, *Christian Counseling and Occultism* (Grand Rapids: Kregel Publications, 1973).

113 Ibid., 22.

114 Robert Pearson, *Storm Over Borneo,* (Overseas Missionary Fellowship, 1967).

115 Rabbi Bahya ben asher, *Biur al ha Toral* (Venice, 1566), 179.

116 Josephus, *The Work of Flavius Josephus; Antiquities of the Jews* (London: G.G. Rutledge), 1.3.1., Book vii.

117 Revue du Musee de Beyrouth.

118 P. J. Wiseman, *New Discoveries in Babylon about Genesis* (London: Masrshall, Morgan and Scott, 2nd Edition), 28-29.

119 Arthur C. Custance, *Genesis and Early Man* (Grand Rapids: Zondervan Publishing House 1975), 103.

120 Ibid., 81ff.

121 J. Cynddylan Jones, *Primeval Revelation* (London: Hodder & Stoughton, 1897), 206-7.

122 Kelly Segraves, *The Great Flying Saucer Myth* (San Diego: Beta Books, 1975), 52.

123 C. S. Dickerson, *Angels Elect and Evil* (Chicago: Moody Press, 1975), 165.

124 John L. Nevius, *Demon Possession and Allied Themes* (Old Tappan, New Jersey: Fleming H. Revell, 1984), 161.

125 Stanton T. Friedman, *UFOs A Complete Nightmare* (Dick Adler, Los Angeles Times, October 20, 1975), 14.

126 Emil Gaverluk, *Did Genesis Man Conquer Space?* (Nashville: Thomas Nelson, Inc., 1974), 88-89.

127 C. S. Dickerson, *Angels Elect and Evil* (Chicago: Moody Press, 1975), 170-174.

128 John A. Kcel, *Strange Creatures from Time and Space* (London: Neville Spearman, 1975), 9.

129 J. Allen Hynek, *The UFO Experience* (Chicago: Henry Regnery Co., 1972), 142.

130 Gordon Creighton, *The Humanoids*, ed. Charles Bowen (Chicago, Henry Regnery Co. 1969), 177.

131 Kelly L. Segraves, *Sons of God Return* (Old Tappan, New Jersey: Fleming H. Revell Co.), 23-24.

132 Basil Tyson, *UFOs: Satanic Terror* (Beaver Lodge, Alberta: Horizon House Publishers, 1977), 67.

133 Gordon Creighton, *The Humanoids*, ed. Charles Bowen (Chicago: Henry Regnery Co., 1969).

134 Ibid., 252.

135 Ibid., 216.

136 Ibid., 243.

137 Ibid., 246-7.

138 Ibid., 149-150.

139 Ibid.,152.

140 Ibid., 59.

141 Ibid., 118.

142 Quoted by Brad Steiger, *Revelation* (Englewood, N.J.: Prentice Hall Inc., 1973), 142.

143 Dr. Clifford Wilson, *UFOs and their Mission Impossible* (New York: Signet Books, 1974), 56.

144 Ibid.,57.

145 Quoted by Brad Steiger, *Revelation,*155.

146 Quoted by John Weldon, *UFOs, What on Earth is Happening?* (Eugene: Harvest House Publishers, 9175), 145-147.

147 Ibid.,145-147.

148 Report by Robert D. Barry, *Christian Beacon,* November 2, 1978.

149 Quoted by Alan and Sally Landsburg, *In Search of Ancient Mysteries* (New York: Bantam Books, 1974), 82.

150 Clifford Wilson, *UFOs and their Mission Impossible* (New York: Signet Books, 1974), 61.

151 Ibid.,70.
152 Gordon Creighton, *The Humanoids*, ed. Charles Bowen (Chicago: Henry Regnery Co., 1969), 192.
153 Ibid., 104.
154 Quoted in *Radar News* (St. Petersburg, Florida) 24, no. 11 (November 1979).
155 Clifford Wilson, *UFOs and their Mission Impossible*, (New York: Signet Books, 1974), 54.
156 Gordon R. Lewis, *Criteria for the Discerning of Spirits*, Demon Possession, ed. John Warwick Montgomery (Minneapolis: Bethany Fellowship, Inc., 1976), 357.
157 Ibid., 357.
158 Gordon Creighton, *The Humanoids*, ed. Charles Bowen (Chicago: Henry Regnery Co., 1969), 71-72.
159 Ibid., 62.
160 Bernard J. Bamberger, *Fallen Angels* (Philadelphia: The Jewish Publication Society of America, 1952), 104.
161 Charles Fort, *Complete Works* (New York: Dover Publications, 1975), 261.
162 Gordon Creighton, *The Humanoids*, ed. Charles Bowen (Chicago: Henry Regnery Co., 1969), 185.
163 Ibid., 249.
164 *National Review*, January 1979.
165 Lewis Sperry Chafer, *Systematic Theology* (Dallas: Dallas Seminary Press, 1947), 2:10.
166 *De Deo Socrates.*
167 Aurelius Augustine, *The City of God*, vol. 1 (Edinburgh: T & T Clark 1872), book VIII, section 16.
168 Kenneth S. Wuest, *Word Studies in the Greek New Testament* (Grand Rapids: Wm. B. Eerdmans Publishing Co., 1966), (1 Peter), 99-100.
169 Emil Gaverluk, *Did Genesis Man Conquer Space?* (Nashville and New York: Thomas Nelson, Inc., 1974) 59.

170 Jacques Vallee, *The Invisible College* (New York: E. P. Dutton, 1975), 233.

171 *Tractatus Theologicus.*

172 Martensen Larsen, *Das Blendwerk des Spiritimus und die Ratsel der seele* (Agentur des Raugen Hauses, 1924), 20.

173 Kurt Koch, *Christian Counseling and Occultism* (Grand Rapids: Kregel Publications, 1973), 165-167.

174 Jacques Vallee, *The Invisible College* (New York: E. P. Dutton, 1975).

175 Robert Southey, *Cornelius Agrippa: A ballad.*

176 John Milton, *Paradise Lost* Complete Poems and Major Prose, ed. Merritt Y. Hughes (New York: Odyssey, 1957).

177 Rousas John Rushdoony, *The Journal of Christian Reconstruction* 1, no. 2 (Winter 1974).

178 C. S. Lewis, "Christianity and Culture," *Christian Reflections,* ed. Walter Hooper (Grand Rapids: Wm. B. Eerdmans Publishing Co., 1967), 33.

179 Jacques Vallee, *Messengers of Deception* (Berkeley: And/Or Press, 1979), 9-10.

180 W. H. Auden, *Danse Macabre* Collected Poems, 1927-1957 (New York: Random House, 1967).

181 Dave Hunt & T. A. McMahon, *The Seduction of Christianity* (Eugene: Harvest House Publishers, 1986), 38.

182 Jesse Penn Lewis, *War on the Saints* (London: Marshall Brothers, 1912), 49.

183 Tertullian, *Apology,* trans. T. R. Glover, Leob Ed. 11,18: XXVII, 6.

184 William Sargent, *The Mind Possessed* (Philadelphia and New York: J. B. Lippincott Co., 1974), 86.

185 Donald Nugent, *Church History* 40 (March 1971): 71.

186 Ken Philpott, *A Manuel of Demonology and the Occult* (Zondervan Publishing House, 1973), 88-90.

187 Isaac Bashevis Singer, *Satan in Goray* (New York: Avon Books, 1963).

188 W. Raymond Drake, *Gods and Spacemen of the Ancient PastThe Dust of Death*

189 Walter Wakefield and Austin P. Evans, eds., *Heresies of the Middle Ages* (New York: Columbia University Press, 1969), 75.

190 Os Guiness, *The Dust of Death* (Downers Grove, Illinois: InterVarsity Press, 1973), 306.

191 Frank Edwards, *Flying Saucers Here and Now* (New York: Tandem Books).

192 Brad Steiger and Joan Whriterow, *Flying Saucers Are Hostile* (New York: Universal Publishing & Distributing Corp., 1967).

193 Gordon Creighton, *The Humanoids,* ed. Charles Bowen (Chicago: Henry Regnery Co., 1969), 204-224.

Made in the USA
Monee, IL
03 November 2026

67396271R00121